Frank Wefers

Partitioned convolution algorithms
for real-time auralization

Logos Verlag Berlin GmbH

λογος

Aachener Beiträge zur Technischen Akustik

Editor:
Prof. Dr. rer. nat. Michael Vorländer
Institute of Technical Acoustics
RWTH Aachen University
52056 Aachen
www.akustik.rwth-aachen.de

Bibliographic information published by the Deutsche Nationalbibliothek

The Deutsche Nationalbibliothek lists this publication in the Deutsche Nationalbibliografie; detailed bibliographic data are available in the Internet at http://dnb.d-nb.de .

D 82 (Diss. RWTH Aachen University, 2014)

ISBN 978-3-8325-3943-6
ISSN 1866-3052
Vol. 20

Logos Verlag Berlin GmbH
Comeniushof, Gubener Str. 47,
D-10243 Berlin
Tel.:   +49 (0)30 / 42 85 10 90
Fax: +49 (0)30 / 42 85 10 92
http://www.logos-verlag.de

PARTITIONED CONVOLUTION ALGORITHMS FOR REAL-TIME AURALIZATION

Von der Fakultät für Elektrotechnik und Informationstechnik der
Rheinischen-Westfälischen Technischen Hochschule Aachen
zur Erlangung des akademischen Grades eines
DOKTORS DER NATURWISSENSCHAFTEN
genehmigte Dissertation

vorgelegt von

Dipl.-Inform.
Frank Wefers
aus Neuss

Berichter:
Universitätsprofessor Dr. rer. nat. Michael Vorländer
Universitätsprofessor D.Sc. Lauri Savioja

Tag der mündlichen Prüfung: 25. September 2014

Diese Dissertation ist auf den Internetseiten der Hochschulbibliothek online verfügbar.

Abstract

Virtual Reality (VR) aims at the creation of responsive simulations, that provide humans the illusion of a world or environment, they can interact with. Therefore, the user is stimulated with sensory cues that are computer-generated, based on a model of a virtual world (scene). Considering the sense of hearing, the acoustic description of the scene is transformed into auditory stimuli, which are then provided using headphones or loudspeakers. Signal processing is fundamental to this process, called auralization. It involves digital filtering in several uses and in diverse forms (e.g. non-linear and linear filtering, time-invariant and time-varying filtering). A common requirement for VR is a low latency (immediate system response). The computational extent however, ranges from moderate to highly complex, depending on the application.

This work focuses on finite impulse response filters (FIR filters), which are applied in binaural synthesis, spatial sound reproduction and artificial reverberation. Straightforward FIR filtering in the time-domain fails to satisfy the requirements stated above. These are met by implementing the FIR filtering using efficient mathematical algorithms for fast convolution. Since the 1960s different algorithmic concepts have been developed, often from the divide-and-conquer paradigm. The most popular example is fast convolution using the fast Fourier transform (FFT), which established as the standard tool. However, also fast convolution algorithms must be adapted to serve for real-time filtering. The most powerful concept hereby is partitioned convolution, which first splits the operands and then solves the partial problems using a fast convolution technique. Essential is that the decomposition conforms with real-time processing.

This thesis considers three different classes of partitioned convolution algorithms for the use in real-time auralization: uniformly and non-uniformly partitioned filters, as well as unpartitioned filters. The algorithmic properties of each class are derived and guidelines for an optimal choice of parameters are provided. All techniques are analyzed regarding multi-channel processing, networks of filters and time-varying filtering, as needed in Virtual Reality. The work identifies suitable convolution techniques for different applications, ranging from resource-aware auralization on mobile devices to extensive room acoustical auralization on dedicated multi-processor systems.

Zusammenfassung

Virtuelle Realität (VR) schafft eine künstliche Wirklichkeit, in die ein Mensch eintauchen und mit der er interagieren kann. Ausgehend von der Beschreibung einer virtuellen Szene werden hierzu mit Hilfe von Computern, verschiedene Sinnesreize generiert, welche beim Benutzer die Illusion der Präsenz in dieser Wirklichkeit erzeugen. Für den Hörsinn bedeutet dies, dass man die akustische Beschreibung einer Szene in hörbare Signale überführen muss, welche dem Benutzer entsprechend dargeboten werden. Für diesen Prozess, der Auralisierung genannt wird, sind digitale Filter ein grundlegendes Werkzeug, das in verschiedenen Arten benötigt wird (z.B. lineare/nichtlineare Filter, zeitinvariante/zeitveränderliche Filter). Eine in der VR allgemeine Anforderung sind geringe Latenzen (möglichst zeitnahe Reaktion). Der Rechenaufwand hierfür reicht, je nach Anwendung, von moderat bis hochkomplex.

Diese Arbeit befasst sich mit digitalen Filtern, welche endliche Impulsantworten haben, sogenannte FIR-Filter (von engl. *finite impulse response*). Für diese finden sich zahlreiche Anwendungen in der akustischen virtuellen Realität, wie beispielsweise in der binauralen Sythese, räumlichen Klangwiedergabeverfahren und bei der Erzeugung künstlichen Nachhalls. finite impulse response (FIR)-Filter können auf einfache Weise im Zeitbereich implementiert werden. Diese Art der Realisierung erfordert allerdings einen erheblichen Rechenaufwand und scheidet dadurch für die oben genannten Anwendungen aus. FIR-Filter können mit Hilfe schneller Faltungsalgorithmen effizienter implementiert werden.

Seit Anfang der 1960er Jahre wurden verschiedene Konzepte zur schnellen Faltung entwickelt, häufig ausgehend vom "teile und herrsche" (divide-and-conquer) Paradigma. Das populärste Beispiel hierfür ist die schnelle Faltung mittels der schnellen Fouriertransformation (engl. fast Fourier transform, FFT), welche sich als Standardverfahren etablierte. Leider sind die meisten schnellen Faltungsverfahren nicht direkt zur Filterung von Signalen in Echtzeit geeignet. Als leistungsfähigstes Konzept hat sich hierbei die Technik der partitionierten Faltung herausgestellt. Dieses zerteilt zunächst die Operanden der Faltung (Partitionierung) und realisiert dann die gewünschte Filterung mittels schneller Faltungen dieser Teilprobleme. Die Art der Zerlegung bestimmt hierbei maßgeblich die Fähigkeit der Echtzeitverarbeitung.

Die vorliegende Arbeit untersucht drei Klassen von partitionierten Faltungen, welche für Echtzeit-Auralisierungen geeignet sind: Algorithmen, welche Filter als Ganzes (d.h. unpartitioniert) verarbeiten und solche, welche Filter in gleiche Teile (uniform) und ungleiche Teile (nicht uniform) zerlegen. Für jede Klasse werden die algorithmischen Eigenschaften im Detail hergeleitet und analysiert und Richtlinien für die optimale Wahl der Parameter werden angegeben. Dabei werden alle Techniken auch hinsichtlich weiterführender Aspekte untersucht, welche für die virtuelle Realität relevant sind, wie Mehrkanal-Filterung, Zusammenschaltungen von Filtern zu Netzwerken, sowie zeitveränderliche Filterung. Die Arbeit identifiziert die geeigneten Faltungstechniken (Filterungsverfahren) für die oben genannten Anwendungen auf verschiedenen Endgeräten, von Auralisierung auf mobilen Endgeräten mit begrenzter Rechenkapazität bis hin zu umfangreichen Raumakustik-Auralisierungen auf speziellen Multiprozessorsystemen.

Contents

Notation and symbols

Elementary math

$i,\ j,\ k,\ m,\ n$	Indices and superscripts
$(a, b, c, \dots)$	Tuple with elements $a, b, c, \dots$
$\vec{u},\ \vec{v},\ \vec{w},$	Vectors
$C,\ F,\ H$	Matrices
$\mathrm{diag}(\dots)$	Diagonal matrix
$\vec{u}^{\mathsf{T}},\ C^{\mathsf{T}}$	Transpose of a vector or matrix
$X(z) = x_0 + x_1 z + x_2 z^2 + \dots$	Polynomial over variable z
$\deg(\cdot)$	Degree of a polynomial
$\mathrm{e},\ \mathrm{e}^x$	Euler constant, exponential function
$\log$	Natural logarithm (base e)
$\log_2$	Logarithmus dualis (base 2)
$i = \sqrt{-1}$	Imaginary unit
$\overline{a + bi} = a - bi$	Complex-conjugate
$\Re\{\cdot\},\ \Im\{\cdot\}$	Real, imaginary part of a complex number
$W_N = e^{-2\pi i/N}$	Primitive N^{th} roots of unity in $\mathbb{C}$
$\Omega(\cdot),\ \mathcal{O}(\cdot)$	Asymptotic lower and upper bound (Landau notation)

Discrete signals and spectra

$x(n) = [x(0), x(1), \dots]$	Discrete-time signal (finite or infinite length)
$X(k) = [X(0), \dots, X(K{-}1)]$	Discrete spectrum (finite length K)
$x(i), X(i)$	Sample or spectral coefficient of index i
$x_i(n)$	Block of index i in the signal $x(n)$ (continuous audio streams)
$x^{(i)}(n)$	i^{th} input or output signal (multichannel)
$h_i(n)$	Sub filter of index i (filter partitions)
$h^{(i)}(n)$	i^{th} filter in an assembly
$\tilde{x}(n) = x\langle n\rangle_N$	N-periodic continuation of $x(n)$
$\delta(n)$	Unit impulse

Operators

$\mathcal{T}\{\cdot\}$, $\mathcal{T}^{-1}\{\cdot\}$	Transform, inverse transform
$\mathcal{DFT}_{(N)}\{\cdot\}$	N-point discrete Fourier transform
$\mathcal{DFT}_{(N)}^{-1}\{\cdot\}$	N-point inverse discrete Fourier transform
$\mathcal{P}_N\{\,x(n)\,\}$	Right-side zero-padding of $x(n)$ to length N
$\mathcal{R}_N\{\,x(n)\,\}$	Rectangular window, extracting the first N elements of $x(n)$
○——●	Transformation time $\rightarrow$ frequency domain
●——○	Transformation frequency $\rightarrow$ time domain
$\times$	Pairwise or element-wise multiplication
$*$, $\circledast$, $\widetilde{\circledast}$	Linear, circular and symmetric convolution

Number theory

$\lfloor\cdot\rfloor$, $\lceil\cdot\rceil$	Floor and ceiling function
$a \mid b$, $a \nmid b$	a divides b, a does not divide b
$a \equiv b \mod N$	Congruence relation
$\langle n\rangle_N$	Integer n modulo N (Rader notation)
$\gcd(a, b)$	Greatest common divisor
$\text{lcm}(a, b)$	Least common multiple

Algebraic structures

$\mathbb{N}$, $\mathbb{Z}$	Natural numbers, integers
$\mathbb{R}$, $\mathbb{C}$	Real numbers, complex numbers
$\mathbb{N}_0 = \mathbb{N} \cup \{0\}$	Natural numbers with zero
$\mathbb{Z}_M = \mathbb{Z}/M\mathbb{Z} = \{0, \ldots, M-1\}$	Set of integers modulo M
R, K	Ring, field
$R[z]$	Ring of polynomials (over variable z)
K^M	M-element vector space over the field K
$K^{M\times N}$	Set of $M \times N$-matrices over the field K

Remarks

Unless outlined, indices begin with 0
Polynomials are defined over the variable z

Acronyms

AVR	Acoustic Virtual Reality
C2R	Complex-to-real
CCP	Cyclic convolution property
CCS	Complex-conjugate symmetric
CFM	Common-factor map
CMAC	Complex-valued multiply-accumulate
CMP	Convolution multiplication property
CMUL	Complex-valued multiply
CPU	Central processing unit
CRT	Chinese remainder theorem
DAG	Directed acyclic graph
DCT	Discrete cosine transform
DFT	Discrete Fourier transform
DHT	Discrete Hartley transform
DIF	Decimation-in-frequency
DIT	Decimation-in-time
DP	Dynamic programming
DSP	Digital signal processor
DST	Discrete sine transform
DTT	Discrete trigonometric transform
FDL	Frequency-domain delay-line
FDN	Feedback delay network
FFT	Fast Fourier transform
FHT	Fast Hartley transform
FIR	Finite impulse response
FNT	Fermat number transform
GA	Geometrical acoustics
GPU	Graphics processing unit
GUPOLS	Generalized uniformly-partitioned Overlap-Save
HRIR	Head-related impulse response
HRTF	Head-related transfer function
IDFT	Inverse discrete Fourier transform
IFFT	Inverse fast Fourier transform
IIR	Infinite impulse response
LTI	Linear time-invariant
MDCT	Modified discrete cosine transform

MDF	Multidelay block frequency domain adaptive filter
MIMO	Multiple-input multiple-output
MISO	Multiple-input single-output
MNT	Mersenne number transform
NTT	Number theoretic transform
NUPOLA	Non-uniformly partitioned Overlap-Add
NUPOLS	Non-uniformly partitioned Overlap-Save
OLA	Overlap-Add
OLS	Overlap-Save
OS	Operating system
PC	Personal computer
PFA	Prime-factor algorithm
PFM	Prime-factor map
R2C	Real-to-complex
RIR	Room impulse response
SIMD	Single-instruction multiple-data
SIMO	Single-input multiple-output
SISO	Single-input single-output
SPS	Symmetric periodic sequence
STFT	Short-time Fourier transform
TDL	Tapped delay-line
TDM	Time-division multiplexing
TSC	Time-stamp counter
UPOLA	Uniformly partitioned Overlap-Add
UPOLS	Uniformly partitioned Overlap-Save
VDL	Variable delay-line
VR	Virtual Reality
WFTA	Winograd Fourier transform algorithm

1. Introduction

Acoustic Virtual Reality (AVR) aims at the simulation of the acoustics in a non-existent world by the help of computers. The user is provided with generated auditory cues, that shall give him a feeling of presence in the virtual environment (immersion). A fundamental necessity for the belief in the simulation is that it confirms with the laws of physics and allows for interaction. The content of virtual scenes can emerge from reality (e.g. acoustic in cars, in traffic, in architecture, musical performances, etc.) or be fictional (e.g. computer games). Virtual reality has many uses. It can blend into our daily life and support us (e.g. 3D telephone). Advances in simulation technique make AVR nowadays usable for assessment tasks (e.g. noise scenarios) and planning tasks (e.g. acoustic in rooms, buildings, spaces).

A central terminus in acoustic virtual reality is *auralization* [56]. It covers all necessary steps to create audible sound (stimuli) from the abstract description of a virtual world (scene). Auralization covers many partial aspects: synthesis (artificial generation of sound), simulation (sound field in an environment), rendering (applying the simulated sound field parameters to the audio signals) and reproduction (presentation of the generated stimuli to the user). Signal processing is fundamental to all of them. The basic tool for modifying the sounds to the need of the applications are digital filters of a variety of types. An example for non-linear filters are variable delaylines (VDLs), used to simulate time-varying propagation delays (Doppler shifts) [106]. Linear filters are used in form of filter banks (e.g. directivity of sound sources, medium attenuation, transmission modeling), for head-related transfer functions (HRTFs) in binaural technology and for simulation reverberation using room impulse responses (RIRs)

Linear filters are divided in two different classes: Feed-forward filters with finite impulse responses (FIR filters) and filters which facilitate feedback loops, resulting in potentially infinite impulse responses (IIR filters). Both types of filters are widely used in auralization. FIR filters allow complete control over the filter characteristic and avoid instabilities by concept. Unfortunately, they can demand a high computational effort (large number of arithmetic operations). The utilization of feedback loops in infinite impulse response (IIR) filters allows a significant reduction of the effort. However, their design is not trivial and issues of stability must be considered.

1

The computational burden of finite impulse response (FIR) filters is overcome by facilitating the tools of mathematics and implementing them with *fast convolution methods*. Their history dates back to the 1960s. Fast convolution techniques are manifold and have been developed from quite diverse mathematical fields. These include classical and linear algebra and number theory. Unfortunately, most of fast convolution algorithms can not be directly applied to real-time filtering and must be adapted accordingly. This is due to the fact, that real-time processing requires partial results to be provided during the convolution. Moreover, they are not computationally efficient, when short blocks of the signal are convolved with long impulse responses. *Partitioned convolution* solves this issue by decomposing large convolutions into better manageable shorter convolutions, while preserving a low latency. This makes it an essential algorithmic tool for the design of efficient real-time FIR filters.

1.1. Objective

This thesis researches how FIR filters can be realized by partitioned convolution for applications in acoustic virtual reality. The filtering tasks in this field are characterized by a variety of requirements. The objective of this work is to identify and examine suited algorithms for these tasks. Immediate system responses are a fundamental requirement for interactive virtual environments. Hence, the focus lies on real-time FIR filters, which process the audio signals with minimal delays (latencies). A main intend is to realize the filters with the least computational load, the least possible runtimes or, in other words, within a minimal number of processor cycles. From a theoretical point of view, the complexity of algorithms can be expressed by the number of operations (mainly arithmetic operations, like multiplications and additions). On practical machines however, the performance of digital signal processor (DSP) algorithms is strongly affect by many further aspects, like the memory access, the caching and branching behaviour as well as the parallelization capabilities. In order to achieve an optimal performance, these aspects must be considered likewise.

This thesis regards convolution algorithms from both perspectives: their analytic complexity *and* their performance on actual hardware. A frequently used term in this work is the computational efficiency of an algorithm. A high efficiency is achieved, when a particular filtering task is accomplished with a comparably low number of operations, regarding the problem from the analytical point of view, or, considering the practical perspective, within the least number of processor cycles. Both definitions asks of course for a reference, which is found in the conceptually simple, but computationally expensive time-domain FIR filters. A high efficiency is needed to overcome the present computational bottleneck, which limits the maximal number of sound sources in a virtual environment. In high-performance computing, simulations of more sources in very complex environments become possible. On mobile devices, auralization becomes less energy consuming. Here, a high computational efficiency helps saving battery life. A fast and efficient implementation is not the only attribute of interest. Section 1.3 outlines several further properties which are of interest in target applications.

1.2. Related work

Auralization has been an area of intensive research. The foundation for many simulation techniques can be traced back to the 1960s. The rising of computers allowed substantial advances in many fields of science and paved the way towards the above stated technologies. This accords in particular for the simulation of acoustics (physics), digital signal processing and computer science. Some notable advances are briefly summarized in the following.

The first approaches in simulating the sound fields in rooms date back to the end of the 1960s. At time, the very limited capabilities of computers and electronics did not allow for simulations based on the fundamental physical descriptions, the wave equations. Instead, the paradigm of geometrical acoustics (GA) was conceived and in 1968, Krokstad [58] introduced the ray-tracing technique for simulating sound fields in rooms. By the end of the 1970s, Allen and Berkley [6] published the image source technique which enabled a precise computation of the early reflections in a room. During the 1980s, the first sound field simulations based on the fundamental physical descriptions, the wave equations, were realized. Smith [97] considered digital waveguide networks for reverberation. Vorländer [117] combined both GA approaches, the image source method and ray-tracing into an efficient hybrid simulation technique. These techniques were later refined (e.g. by hierarchical search structures) and became efficient enough to simulate the sound field in rooms in real-time. Botteldooren [15] applied the finite-difference time-domain for wave-based simulations of room acoustics. While GA established itself as a standard tool for engineers and scientists, recent advances in computing hardware brought wave-based simulations into the realm of real-time simulations. This became mainly possible, due to very powerful graphic processors (GPUs), which can be successfully applied to solve acoustic problems as well [113, 86, 118, 70, 82]. Savioja [84] achieved wave-based simulations of the lower frequency sound field in real-time on these devices.

Sound field simulations are one fundamental aspect of auralization. Digital filtering marks another cornerstone. The line of developments in simulation techniques was accompanied by significant advances in digital signal processing, particularly fast and efficient filtering techniques. In the 1960s, Schröder

3

and Logan [92, 91] layed the theoretical foundation for artificial reverbera-
tor networks. Aware of the limited hardware capabilities, they emulated
the reverberation in a room by low-complexity IIR filter networks consisting of all-passes, comb filters and delay-lines. In 1965, Cooley and Tukey
[24] published their fast Fourier transform (FFT) algorithm, which helped to
establish the discrete Fourier transform (DFT) as the common tool it nowadays is. Shortly after, Stockham [105] outlined the use of the FFT for fast
convolution and correlation. This marked a milestone for the development
of efficient FIR filtering algorithms. In the 1980s, uniformly-partitioned convolution was presented by Kulp [59]. Meanwhile, Schröder's and Logan's
original IIR reverberators were developed further, leading to comprehensive
feedback delay networks (FDNs) [52]. At the beginning of the 1990s, the
hardware became fast enough so that many applications in the field of virtual acoustics approached the realm of possibility. Hardware-accelerated FIR
filters enabled the first real-time auditory environments with reverberation
[32]. Convolvotron was a PC extension card equipped with a large number
of parallel DSPs. It could binaurally synthesize up to eight free-field sound
sources in real-time. Alternatively, a limited number of early reflection was
possible. The FFT computation advanced by the (re)discovery of the split-
radix FFT algorithm by Duhamel and Vetterli [29]. The concept of non-
uniformly partitioned convolution was proposed by Egelmeers and Sommen
[30] and popularized by Gardner [39]. Gardner and McGrath [80] considered FIR filtering using fast convolution for simulating reverberation. The
Huron digital audio workstation [69] implemented non-uniformly partitioned
convolution using DSPs and pushed the boundary of large-scale FIR filtering
a large leap forward. Multi-channel convolutions with several 100,000 filter
coefficients became possible at a low latency.

Interactive auralizations in real-time and acoustic virtual reality systems
started to appear towards the end of the 1990s in scientific groups around
the world: the DIVA virtual audio reality system ([46, 85] developed at Aalto
University, Finland (former Helsinki University of Technology), the sound lab
system (abbreviated SLAB) developed at NASA Ames Research Center [125],
the virtual reality system at RWTH Aachen University ([63, 64, 89, 120]), the
REVES research project at INRIA in Sophia-Antipolis [114, 104] and at the
University of North Carolina at Chapel Hill [108]. Recent systems are evolv-
ing towards the simulation of outdoor scenarios [70]. As research in acoustic
virtual reality continues, the level of detail in simulations keeps increasing.
Lately, acoustic virtual reality approaches mobile devices [55, 83]. Nevertheless, after 50 years of research, the simulation of acoustics in real-time
still marks an extraordinary challenge and several questions are unanswered
yet—not excluding signal processing.

1.3. Problem description

This work targets the efficient realization of real-time FIR filters using general-purpose processors. Hence, the considerations are limited to a block-based audio processing (cp. Sec. A.1). A real-time FIR filtering problem is defined as follows: A continuous audio stream is processed in length-B blocks. B is referred to as the *block length*. The terms '*streaming buffer size*' and '*frame size*' are often used alternatively in audio software development. The sampling rate of the audio stream is f_S. The resulting frame rate (processed frames per second) is given by $R = f_S/B$. Unless outlined, a single channel is considered only. Figure 1.1(a) shows the corresponding block diagram. The input signal $x(n)$ consists of consecutive length-B blocks $x_0(n), x_1(n), \cdots$ (sub indices denote the individual blocks). It is filtered with finite impulse response $h(n)$ of N filter coefficients. The filtering is processed *block by block*. When the ith input block $x_i(n)$ is fed into the filter, the filtering algorithm computes the corresponding output block $y_i(n)$. The computation time is bounded by $T = B/f_S = 1/R$, the duration of one frame. If this time budget is exceeded, the audio stream is interrupted and dropouts are the consequence. Precautions must be taken in order to prevent this. Therefore, a certain safety margin is incorporated and the time budget is not fully exploited. Several subsequent considerations require a clear definition of the audio processing procedure, in particular its events and their timescale. This is given in A.1 in the appendix.

The above stated problem considers a single input and single output only (SISO system). The problem can be generalized to multiple inputs and multiple outputs (MIMO system), interconnected by intermediary FIR filters $h_{i \to j}(n)$. This is illustrated in figure 1.1(c). All inputs and outputs request and provide blocks of the same length. The admissible time span for computations T is the same as above. Signal processing for auralization often incorporates assemblies of individual filters in serial (figure 1.1(b)) or parallel (figure 1.1(d)). These occur as cascaded filters on a sound path (e.g. directivity, medium attenuation, etc.) or at points of superposition (e.g. the listeners ears, coupling joints between separated spaces).

The regarded FIR filters are linear and time-invariant systems (LTI systems). Interaction in virtual environments changes the auralization filters over time. Time-varying FIR filtering has the following meaning in this thesis: at a distinct point in time, a current filter impulse response $h_0(n)$ is replaced with a new filter $h_1(n)$. Strictly speaking, both individual filters are still LTI systems. The exchange $h_0(n) \to h_1(n)$ can usually not be accomplished by instantaneous switching of all coefficients within a single sample. This mostly causes audible artifacts. Hence a smooth transition is realized over a number of L output samples. Even within this period, both sets of filter coefficients remain constant. The transition is achieved by crossfading the outputs of both filters.

Interaction demands imperceptibly short response times of a virtual reality system. With respect to the filtering, two different types of latency occur: Input-to-output latency is the duration between the events of exciting the filter with a signal and receiving a response at its output. It is adjusted by selecting a reasonably short audio processing block length B. Frequency-domain techniques require the impulse response to be transformed, before it can be used. This transformation consumes time. Moreover, the exchange itself can be bound to specific points in time, which introduces further waiting times. The filter exchange latency describes the delay, when the update of a filter is initiated, until it is exchanged and affects the output of the filter.

1.4. Outline

The thesis is organized as follows: Chapter 2 reviews a wide variety of fast convolution methods and assesses their use for real-time filtering. The objective is to identify the most promising base technology for partitioned convolution methods. The chapter's intention is also to scrutinize FFT-based fast convolution in its status as a standard method. Several divide-and-conquer strategies are examined for their real-time compliance. Chapter 3 introduces the fundamentals of partitioned convolution. Partitions of both operands— signal and filter—are formally defined. Common processing strategies are reviewed. A classification of partitioned convolution techniques is presented. The remaining part of the thesis is dedicated to the study of three of these classes, which are useful for real-time FIR filtering. The subsequent chapter 4 reviews basic and straightforward techniques of real-time filtering using the FFT. These techniques have in common that they do not partition the filter impulse response. The examination of these algorithms aims at the identification of their weaknesses and as proof for the importance of higher-level techniques, which partition the filters as well. Further aspects like filters with multiple inputs and multiple outputs (MIMO filters), assemblies of filters and the implementation of a time-varying filtering are firstly developed for these conceptually simple techniques. Chapter 5 and 6 consider methods with filter partitioning. The state-of-the-art algorithms are presented and their properties are examined in detail. This includes the choice of parameters, their dependencies, the algorithms' performance and their runtime complexity classes. Towards the end of both chapters the methods are reflected with the advanced aspects, stated above. The benchmark procedure, test system and its performance data is described in depth in chapter 7. Finally, the findings of this thesis are summarized in chapter 8. Some guidelines on the choice of algorithms and parameters are provided. Open scientific questions

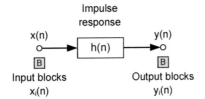

(a) FIR filter (single channel)

(b) Cascade of FIR filters

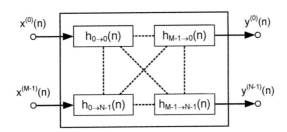

(c) FIR filter with multiple inputs and outputs

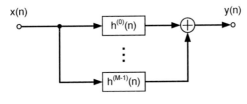

(d) Parallel FIR filters

Figure 1.1.: Types of FIR filters and assemblies

are quoted in the outlook. A description of real-time audio processing, comprehensive lists of results and additional mathematical correspondences are found in the appendix.

1.5. Contributions

The main contributions of this thesis are:

- Review of fast convolution algorithms for real-time filtering and identification of the most suitable methods on general purpose processors.
- Classification of partitioned convolution algorithms. The central part of the thesis is dedicated to the examination of the three classes used for real-time filtering: methods with a non-uniform, uniform and no filter partitioning.
- Introduction of benchmark-based semi-empirical cost models of the algorithms, allowing to capture and account for specific properties of a target machine.
- Examination of uncommon FFT transform sizes (i.e. non powers-of-two) for real-time filtering.
- Development of a generalized uniformly-partitioned convolution technique, featuring individual partitions in both operands, input signal and filter. Viable parameters and potential computational benefits are examined.
- Considerations on time-varying filtering in conjunction with frequency-domain convolution techniques. Introduction of a computationally efficient formulation of crossfading in the DFT domain.
- Considerations on the efficient implementation of MIMO filters with all regarded running convolution techniques.
- Considerations on the efficient implementation of sequential and parallel assemblies of frequency-domain filters.
- Formal derivation of general timing dependencies in non-uniformly partitioned convolution techniques, also respecting individual sub filter iterations.
- Derivation of the runtime complexities of all regarded real-time filtering techniques. Comparison of their computational costs for different filter lengths and latency requirements (block lengths).
- Guidelines for the choice of algorithms and selection of parameters for real-time FIR filtering.

2. Fast convolution techniques

This chapter gives an overview of fast convolution algorithms and their historic evolution. All techniques are reviewed with respect to the previously enumerated objectives and requirements (see section 1.3). The objective of this chapter is to identify the algorithms with the least computational complexity and usability for real-time filtering techniques, which are subject of the subsequent chapters.

A fundamental design technique for fast convolution algorithms is to express convolution operations with the concepts and tools of another mathematical field. This facilitates it to apply methods from this field to the original problem. All fast convolution methods have origins in linear algebra, polynomial algebra and number theory or combine techniques from these fields. Matrix diagonalization in linear algebra is a corner stone of *transform-based fast convolution* algorithms–like FFT-based fast convolution. This class marks the most important class and is reviewed in the most detail in this chapter. Another technique known as *interpolation-based convolution* has its roots in classical algebra. It makes use of polynomial interpolation to derive simplified equations for discrete convolutions, with fewer terms than the discrete convolution sum. Last but not least did *number theory* lead to many, often revolutionary, new approaches. Many of these techniques were conceived in the area of fast Fourier transform algorithms and then later applied to convolution methods as well. The central concept thereby have become factorial rings and fields (e.g. calculations modulo integers or polynomials). Many techniques involve the chinese remainder theorem (CRT).

In the almost 60 years that passed since the famous paper by Cooley and Tukey [24] had been published in 1965, the fast Fourier transform certainly became very popular and probably the most important tool in digital signal processing. Today, most people associate *fast convolution* with FFT-based convolution algorithms. And there are several good arguments, why FFT-based convolution is an excellent choice. For instance, the intensive research that has been done on fast algorithms and the availability of very matured high-performance libraries. However, several other approaches to fast convolution have been researched as well. Nowadays, many of these techniques are overshadowed by the enormous success of FFT convolution. FFT-based fast convolution can be thought of as the *reference method*—not only in this thesis, but also in practice. This chapter scrutinizes this statement and therefore,

9

carefully analyzes and compares the algorithmic complexity of very different fast convolution methods. It poses the simple, yet fundamental question: Is FFT-based convolution nowadays the most reasonable technique for real-time FIR filtering on general-purpose processors?

The chapter is organized as follows: In the beginning the elementary operations of linear and circular convolution are reconsidered and their relations to the different mathematical areas are shown up. The remaining part of the chapter is dedicated to the review of the three important classes of algorithms, which were enumerated above. Emphasis lies on the most important techniques, in particular FFT-based convolution, but a comprehensive overview of the field is aimed. Most of the methods discussed here, were initially not designed for real-time processing. It is evaluated how they can be adapted to suit this purpose. If possible, the computational complexity is reviewed. The chapter ends with a summary of the methods.

2.1. Discrete convolution

Discrete convolution is an operator for two sequences $x(n)$ and $h(n)$ of indefinite lengths. It is defined by the well-known formula [74]

$$y(n) = x(n) * h(n) = \sum_{k=-\infty}^{+\infty} x(k) \cdot h(n-k) \qquad (2.1)$$

$$= h(n) * x(n) = \sum_{k=-\infty}^{+\infty} h(k) \cdot x(n-k)$$

The result of this convolution operation $x(n) * h(n)$ is a sequence $y(n)$, also of indefinite length. In signal processing the sequences are referred to as *signals*.

Discrete convolution is of fundamental importance in system theory. Given a linear and time-invariant system (LTI system), which is fully described by its impulse response $h(n)$, the discrete convolution in Eq. 2.1 defines the

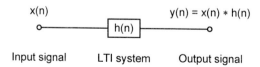

Input signal   LTI system   Output signal

Figure 2.1.: Correspondences between the input signal, the output signal and the filter impulse response of a linear time-invariant system

resulting signal $y(n)$ at the output of the system, when a discrete-time signal $x(n)$ is given into its input (figure 2.1). $h(n)$ is the filter impulse response or short *filter*. Signals of indefinite length are needed for theoretical analysis. In practical applications on computers, at least the filter $h(n)$ has a definite length (finite impulse response). The domain K of sample values $x(n), y(n)$ and filter coefficients $h(n)$ can be integers ($\mathbb{Z}$), real numbers ($\mathbb{R}$) or complex numbers ($\mathbb{C}$) in floating representation.

Two different cases of discrete convolution can be distinguished in Eq. 2.1

- The convolution of two sequences which *both have finite lengths*. This marks the most general convolution operation on computers. It is the standard operation for *offline* audio processing, e.g. filtering an audio file with a specific finite impulse response.

- The convolution of *infinite-length* sequence with a *finite-length* sequence, often called *running convolution*. A typical application is FIR filtering of a (potentially) infinite stream of audio samples $x(n)$ with a filter impulse response $h(n)$ of a finite length N. This is a typical real-time application, where the output samples $y(n)$ are continuously computed from the input samples $x(n)$ with only a small amount of latency. On computers this is done by means of *blocks* (or *frames*) of a specific *block length* B.

The operator in Eq. 2.1 is *linear convolution*. The relations in Fig. 2.1 are only fulfilled by this operator, making it the desired type of convolution in audio filtering applications. Further discrete convolution operations are known, for instance *circular convolution* (Sec. 2.3), *symmetric convolution* or *skew-symmetric convolution* (Sec. 2.5.6). These operators have favorable mathematical properties. Very often, they are used to realize the desired linear convolution (Sec. 2.3 and Sec. 2.5.6).

2.2. Linear convolution

The linear convolution (symbolized by $*$) of a potentially *infinite-length* sequence $x(n)$ with a length-N filter $h(n) = h(0), \ldots, h(N-1)$ is defined as [73, 74]

$$y(n) = x(n) * h(n) = \sum_{k=0}^{N-1} x(n-k) \cdot h(k) \qquad (2.2)$$

The resulting sequence $y(n)$ is a superposition of shifted versions $x(n-k)$ of the original sequence $x(n)$, weighted by the filter coefficients $h(k)$.

In case that both sequences $x(n) = x(0), \ldots, x(M-1)$ and $h(n) = h(0), \ldots, h(N-1)$ have *finite lengths* $M, N \in \mathbb{N}$, the summation in Eq. 2.2 is limited to indices k for which both sequences, the shifted $x(n - k)$ and $h(k)$, overlap in at least one element $(0 \leq n - k \leq M - 1 \wedge 0 \leq k \leq N - 1)$. Outside these index intervals the values $x(n - k)$ and $y(k)$ are undefined. This yields the definition of linear convolution of two *finite-length* sequences

$$y(n) = x(n) * h(n) = \sum_{k=\max\{0,n\}}^{\min\{n-M+1,N-1\}} x(n - k) \cdot h(k) \qquad (2.3)$$

From the necessary overlapping of the sequences $x(n-k)$ and $h(k)$ it follows, that the output sequence $y(n)$ has the length $M + N - 1$.

Matrix-vector product formulation

Linear convolution in Eq. 2.2 can be interpreted as a matrix-vector product of the form $\vec{y} = H\vec{x}$ in Eq. 2.4. Burrus and Parks provide detailed explanations on the relations between convolution algorithms and matrix operations in their textbook [21] and give several examples. Let in the following K be a field. K^M denotes the M-element vector space over the field K. $K^{M \times N}$ is the set of all $M \times N$ matrices defined over K.

The sequence $x(n)$ corresponds to a M-element vector $\vec{x} = [x_0 \cdots x_{M-1}]^\top \in K^M$ and $y(n)$ to a vector $\vec{y} = [y_0 \cdots y_{M+N-2}]^\top \in K^{M+N-1}$ with $M + N - 1$ elements accordingly. $H \in K^{M+N-1 \times M}$ is a convolution matrix with $M + N - 1$ rows and M columns. Its columns contain shifted versions of the sequence $h(n)$, padded by zeros.

$$
\begin{bmatrix} y_0 \\ y_1 \\ \vdots \\ y_{M+N-2} \end{bmatrix}
=
\begin{bmatrix}
h_0 & 0 & \cdots & 0 & 0 \\
h_1 & h_0 & \cdots & \vdots & \vdots \\
h_2 & h_1 & \cdots & 0 & 0 \\
\vdots & h_2 & \cdots & h_0 & 0 \\
h_{N-2} & \vdots & \cdots & h_1 & h_0 \\
h_{N-1} & h_{N-2} & \ddots & \vdots & h_1 \\
0 & h_{N-1} & \cdots & h_{N-3} & \vdots \\
0 & 0 & \cdots & h_{N-2} & h_{N-3} \\
\vdots & \vdots & \vdots & h_{N-1} & h_{N-2} \\
0 & 0 & 0 & \cdots & h_{N-1}
\end{bmatrix}
\cdot
\begin{bmatrix} x_0 \\ x_1 \\ \vdots \\ x_{M-1} \end{bmatrix}
\qquad (2.4)
$$

Polynomial product formulation

Linear convolution can also be defined by polynomial products [13]. Sequences are represented by polynomials as algebraic structures. The values $x(0), \ldots, x(M-1)$ and $h(0), \ldots, h(N-1)$ can be interpreted as polynomial coefficients $x_0, \ldots, x_{M-1}$ and $h_0, \ldots, h_{N-1}$ of their two generating polynomials $X, H \in R[z]$ with $\deg(X) = M - 1$ and $\deg(H) = N - 1$. $R[z]$ denotes the ring of polynomials defined over a ring or field R and the variable z.[1]

$$X = \sum_{n=0}^{M-1} x(n)z^n = x(0) + x(1)z + x(2)z^2 + \cdots + x(M-1)z^{M-1} \quad (2.5)$$

$$H = \sum_{n=0}^{N-1} h(n)z^n = h(0) + h(1)z + h(2)z^2 + \cdots + h(N-1)z^{N-1} \quad (2.6)$$

Operations on the sequences, like addition, multiplication and shifting, map to operations within polynomial algebra, for instance polynomial addition and multiplication. The linear convolution of the sequences $x(n)$ and $h(n)$ corresponds to the polynomial product

$$y(n) = x(n) * h(n) \quad \widehat{=} \quad Y = X \cdot H \quad (2.7)$$

The resulting polynomial $Y \in R[z]$ has the degree $\deg(Y) = M + N - 2$

$$Y = \sum_{n=0}^{M+N-2} y_n z^n = y_0 + y_1 z + y_2 z^2 + \cdots + y_{M+N-2} z^{M+N-2} \quad (2.8)$$

$$\text{with} \quad y_k = \sum_{i=0}^{k} x_i h_{k-i} \quad (2.9)$$

Its coefficients $y_0, \ldots, y_{M+N-2}$ are defined by a linear convolution in Eq. 2.9 and they correspond to values of the output sequence $y(n)$.

Computational complexity

The simplest way to compute a linear convolution is to evaluate the Eq. 2.2 for all output samples. Sometimes, this is cited as *direct convolution*. In signal processing, this corresponds to filter the input samples with a tapped delay-line (TDL), as shown in Fig. 2.2. In the following the required number of arithmetic operations is derived.

---

[1] In this work z is favored as a polynomial variable, as x is used for sequences and signals.

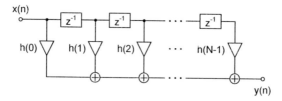

Figure 2.2.: Direct-form FIR filter (tapped delay-line)

Firstly, the running convolution of an infinite input signal with an N-point filter is considered. As the length of the input signal is quasi-infinite, the computational complexity is assessed by the number of arithmetic operations that are necessary to compute *one sample* of filtered output. Evaluating Eq. 2.2 for one output value $y(n)$ requires N multiplications and $N-1$ additions. The number of arithmetic operations $T(N)$ per filtered output sample for the direct running linear convolution with a N-tap filter is

$$T(N) = 2N - 1 \in \mathcal{O}(N) \tag{2.10}$$

Secondly, the number of arithmetic operations $T(M, N)$ for the linear convolution of two sequences with finite lengths M, N is derived from the corresponding N-tap FIR filter (Fig. 2.2). The filtering processing can be split into three phases:

- Within the first $N-1$ steps the accumulators of the filter are getting filled with input samples. Before, they contained zeros. The number of operations in this phase is
$$\sum_{i=1}^{N-1} (2i - 1) = N^2 - 2N + 1.$$

- In the next $M-N+1$ steps all accumulators contain input samples and Eq. 2.10 applies. The number of operations in this phase is $(M-N+1)(2N - 1)$.

- Now all M input samples have been given into the filter and within the next $N-1$ the accumulators fill up with zeros again. Here the number of operations is the same as in the first phase.

The total number of operations is $2(N^2 - 2N + 1) + (M - N + 1)(2N - 1)$. Simplifying this expression, the exact number of arithmetic operations for an $M{\times}N$ direct linear convolution is

$$T(M, N) = 2MN - (M + N - 1) \in \mathcal{O}(M \cdot N) \tag{2.11}$$

Zeros in the accumulators of an FIR filter should not be neglected. A 16×16 linear convolution requires 481 operations. Computing a 16-point running convolution for the same number of $16 + 16 - 1 = 31$ output samples demands $31^2 = 961$ operations.

The time complexity of an $N\times N$ direct linear convolution lies within $\mathcal{O}(N^2)$. Any algorithm that computes the same result in a time complexity lower than $\mathcal{O}(N^2)$ is considered a *fast linear convolution method* in the following.

2.3. Circular convolution

Linear convolution is *aperiodic* (or *non-cyclic*) [73]. Many fast convolution algorithms perform the convolution operation within the domain of some discrete transform (e.g. the DFT), where it can be realized more efficiently. Most of these transforms assume some periodicity of the sequences. The assumption of periodic sequences leads to an adapted formulation of discrete convolution, which is known as cyclic or *circular convolution*.

Let $x(n)$ be a sequence of the finite length $N \in \mathbb{N}$. The *periodic continuation* of $x(n)$ is defined as $\widetilde{x}(n) = x\langle n\rangle_N$, where the indices $-\infty < n < \infty$ are evaluated modulo the period N. $\langle\cdot\rangle_N$ denotes the residual of the integer n modulo N [68]: $\langle n\rangle_N = k \Leftrightarrow n \equiv k \mod N$. $\widetilde{x}(n)$ is periodic in N samples.

The N-point circular convolution (symbolized by $\circledast$) of two length-N sequences $x(n)$ and $h(n)$ is defined as the sum [74]

$$\widetilde{y}(n) = x(n) \circledast h(n) = \sum_{k=0}^{N-1} \widetilde{x}(n-k) \cdot \widetilde{h}(k) = \sum_{k=0}^{N-1} x\langle n - k\rangle_N \cdot h(k) \quad (2.12)$$

As the indices of the reversed and shifted sequence $x\langle n - k\rangle_N$ are evaluated modulo N, the output sequence $\widetilde{y}(n)$ is periodic in every N^{th} element as well. Therefore, it is fully determined by N values of $\widetilde{y}(n)$ and has the length N. As the index k is in the range $0 \leq k < N$ anyway, the modulo operation can be dropped for the term $h(k)$.

Matrix-vector product formulation

The matrix-vector product representation of N-point circular convolution has the form $\vec{\widetilde{y}} = C\vec{x}$. The length-$N$ sequences $x(n), h(n)$ and $\widetilde{y}(n)$ correspond to N-element vectors $\vec{x}, \vec{h}, \vec{\widetilde{y}} \in K^N$. $C = (c_{i,j}) \in K^{N\times N}$ is an $N \times N$ Toeplitz matrix, called *circulant matrix* (or *circular convolution matrix*) [19]. Its elements are $c_{i,j} = h\langle i - j\rangle_N$ (with indices $0 \leq i,j < N$). C has the special property that each row or column vector is a copy of its neighboring row or column vectors, shifted by one element. An example length-5 circular convolution has the form

$$
\begin{bmatrix} \widetilde{y}_0 \\ \widetilde{y}_1 \\ \widetilde{y}_2 \\ \widetilde{y}_3 \\ \widetilde{y}_4 \end{bmatrix} = \begin{bmatrix} h_0 & h_2 & h_1 & h_0 & h_2 \\ h_1 & h_0 & h_2 & h_1 & h_0 \\ h_2 & h_1 & h_0 & h_2 & h_1 \\ h_0 & h_2 & h_1 & h_0 & h_2 \\ h_1 & h_0 & h_2 & h_1 & h_0 \end{bmatrix} \cdot \begin{bmatrix} x_0 \\ x_1 \\ x_2 \\ x_3 \\ x_4 \end{bmatrix}
\tag{2.13}
$$

Where in a linear convolution matrix the column vectors are zero-padded and shifted versions of the sequence $h(n)$, the circulant matrix contains shifted versions of the N-periodic continuations $\widetilde{h}(n)$.

Circulant matrices have an important mathematical property: Their eigenvalues $\lambda_0, \ldots, \lambda_{N-1}$ are linear combinations of the filter coefficients $h(n)$ with powers of primitive N^{th} roots of unity $W_N = e^{-2\pi i/N} \in \mathbb{C}$ in the complex number plane [43]

$$
\lambda_j = \sum_{k=0}^{N-1} h(k) W_N^{jk}
\tag{2.14}
$$

This results in eigenvectors of the form

$$
\vec{v}_j = \frac{1}{\sqrt{N}} \left[1, W_N^j, (W_N^j)^2, \cdots, (W_N^j)^{N-1} \right]^{\mathsf{T}}
\tag{2.15}
$$

Eq. 2.14 corresponds to a N-point DFT of $h(n)$ (cp. Eq. 2.31) and $[\vec{v}_0, \ldots, \vec{v}_{N-1}]$ (Eq. 2.15) defines an eigenvector basis of C. Hence the DFT diagonalizes a circulant matrix, resulting in the cyclic convolution property (CCP). These relations are reviewed in more detail in section 2.5.1.

Polynomial product formulation

Circular convolution can be defined using polynomial products as well. The essential difference lies in the algebraic structure, in which the computation is performed. For linear convolution it is computed in the polynomial ring $R[z]$. In order to obtain circular convolution, the calculation need to be performed in modular arithmetic over the quotient polynomial ring $R[z]/(z^N - 1)$ with a polynomial modulus $z^N - 1$ [13]. All polynomials $P \in R[z]/(z^N - 1)$ have a degree $\deg(P) \leq N$. The circular convolution can be computed directly from the linear convolution polynomial $Y = X \cdot H \in R[z]$ by evaluating Y modulo $z^N - 1$ [13], realizing the fold-back of overlapping samples into the N-point period.

$$
y(n) = x(n) \circledast h(n) \quad \widehat{=} \quad Y \equiv X \cdot H \mod z^N - 1
\tag{2.16}
$$

This is illustrated in the following example ($N = 3$):

$$x = \begin{bmatrix} 2 & -1 & 3 \end{bmatrix}$$
$$h = \begin{bmatrix} 1 & 2 & -1 \end{bmatrix}$$

$$X = 2 - z + 3z^2$$
$$H = 1 + 2z - z^2$$

$$x * h = \begin{bmatrix} 2 & 3 & -1 & 7 & -3 \end{bmatrix}$$

$$X \cdot H = (2 - z + 3z^2) \cdot (1 + 2z - z^2) =$$
$$2 + 3z - z^2 + 7z^3 - 3z^4$$

$$x \circledast h = \begin{bmatrix} 9 & 0 & -1 \end{bmatrix}$$

$$X \cdot H = 2 + 3z - z^2 + 7z^3 - 3z^4 =$$
$$(7 - 3z) \cdot (z^3 - 1) + (9 - z^2)$$
$$X \cdot H \mod z^3 - 1 = 9 - z^2$$

Example 2.1: Convolutions and corresponding polynomial products

Linear convolution using circular convolution

Linear-convolution can be implemented using circular convolution. This makes fast circular convolution algorithms applicable for linear filtering. The relations between both operations are illustrated by the examples in figure 2.3. Let $x(n)$ be a length-M sequence and $h(n)$ a length-N sequence. The result of an $M \times N$ linear convolution has the length $M + N - 1$. The K-point circular convolution equals a $M \times N$ linear convolution, when its period K is sufficiently large, so that the $M + N - 1$ samples do not overlap in time (avoiding *time-aliasing*)

$$\mathcal{R}_{M+N-1}\Big\{ \mathcal{P}_K\{ x(n) \} \circledast_K \mathcal{P}_K\{ h(n) \} \Big\} = x(n) * h(n) \quad \Leftrightarrow \quad (2.17)$$

$$K \geq M + N - 1 \tag{2.18}$$

For completeness, $\mathcal{P}_K$ is a padding operator appending zeros until a length K and $\mathcal{R}_{M+N-1}$ is a rectangular window, cutting out the first $M+N-1$ values of the sequence. Given that condition 2.18 is violated and $K < M+N-1$, it is not guaranteed that the linear convolution results do not overlap and consequently does the equality Eq. 2.17 not hold anymore.

2.4. Interpolation-based fast convolution

The class of methods reviewed first are interpolation-based fast convolution techniques. These are fast algorithms for computing linear convolutions and found on the polynomial product formulation, introduced in section 2.2. Computing the linear convolution $x(n) * h(n)$ translates to the problem of finding the polynomial coefficients $y_0, \ldots, y_{M+N-2}$ of $Y(z)$ (Eq 2.8), as motivated in section 2.2. They can be calculated using direct discrete convolution

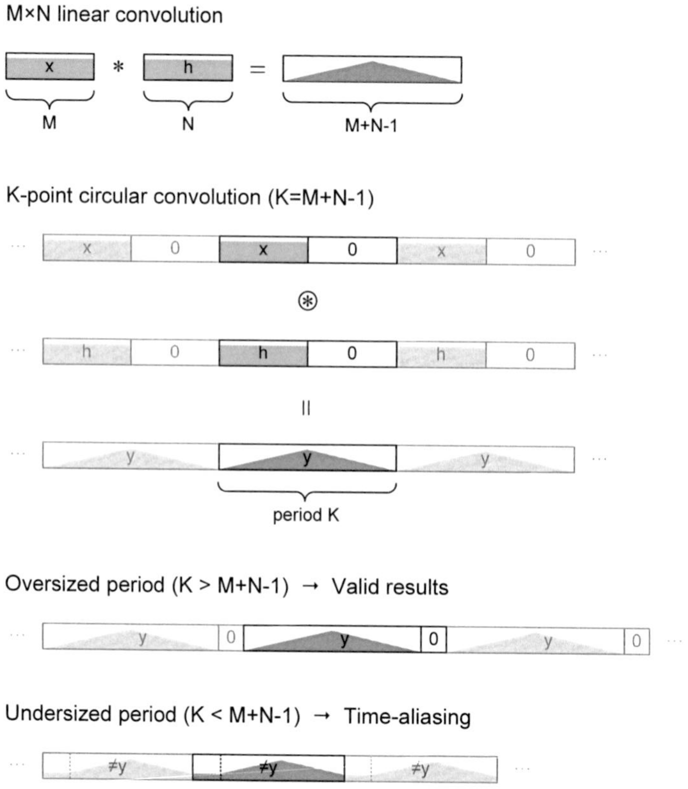

M×N linear convolution

K-point circular convolution (K=M+N-1)

Oversized period (K > M+N-1) → Valid results

Undersized period (K < M+N-1) → Time-aliasing

Figure 2.3.: Realizing linear convolution by circular convolution for two example sequences[2]

(Eq. 2.9), resulting at the runtime complexity of $\mathcal{O}(M \cdot N)$ (cp. section 2.2). A computational advantage arises from transforming the problem of finding the coefficients y_i into an interpolation problem. This allows obtaining formulations (sequences of terms) of linear convolutions that have less arithmetic operations than the naive evaluation. The mathematical framework for interpolation-based convolution is the class of Toom-Cook algorithms. A particular method is the Karatsuba algorithm [54]. It has a runtime complexity $\mathcal{O}(N^{\log_2 3}) \subset \mathcal{O}(N^2)$ and is hence considered *fast* here (cp. section 2.2), although it is asymptotically slower than transform-based approaches in $\mathcal{O}(N \log N)$. Unfortunately, interpolation-based techniques become inefficient for longer filters and suit short convolutions only. Their relevance for

---

[2]Correct scaling is neglected

real-time filtering (also with longer filters), stems from their use in accelerating partition convolutions [47] in conjunction with multi-dimensional index mapping (see section 2.6.1). The index mapping techniques are introduced in section 2.6.1 and their application for real-time FIR filtering is reviewed. For a detailed introduction the reader is referred to the textbook by Blahut [13]. Part of this section is the derivation of the exact number of operations for the Karatsuba convolution algorithm, enabling the comparison to alternative methods.

2.4.1. Toom-Cook algorithm

The Toom-Cook algorithm was originally conceived by Toom [110] in 1963 as a method for fast integer multiplication. Later in 1966, Cook [23] improved the algorithm. Blahut [13] shows how the technique can be used to implement linear convolution. As motivated above, the problem of linear convolution corresponds to finding the coefficients y_k of the product polynomial $Y(z) = X(z) \cdot H(z)$. A fundamental consideration is, that the resulting polynomial $Y(z)$ has $\deg(Y) = M + N - 2$ and is therefore uniquely determined by $M + N - 1$ data points. Instead of explicitly multiplying the *polynomials* $X(z) \cdot H(z)$ using the discrete convolution in Eq. 2.9, the Toom-Cook algorithm evaluates the polynomials $X(z)$ and $H(z)$ for a set of data points α_i and then multiplies their *values* $Y(\alpha_i) = X(\alpha_i) \cdot H(\alpha_i)$. Afterwards the product polynomial $Y(z)$ is constructed using Lagrange interpolation and its coefficients y_k are obtained. The algorithm consists of three phases:

1. $M + N - 1$ distinct supporting points $\alpha_0, \ldots, \alpha_{M+N-2}$ $(\forall i, j : i \neq j \rightarrow \alpha_i \neq \alpha_j)$ are chosen.

2. $X(\alpha_i)$ and $H(\alpha_i)$ are evaluated for all supporting points α_i and multiplied $Y(\alpha_i) = X(\alpha_i) \cdot H(\alpha_i)$.

3. $Y(z)$ is constructed from the $M + N - 1$ data points $(\alpha_i, Y(\alpha_i))$ using Lagrange interpolation

$$Y(z) = \sum_{n=0}^{M+N-2} Y(\alpha_i) L_i(z) \qquad L_i(z) = \prod_{\substack{k=0 \\ k \neq i}}^{M+N-2} \frac{z - \alpha_i}{\alpha_k - \alpha_i} \qquad (2.19)$$

The interpolation approach results in an alternative formulation of linear convolution, in form of a sequence of terms. These terms group common expressions and remove redundant computations, resulting in less operations than the corresponding direct convolution. Hence, the Toom-Cook method allows deriving fast algorithms for short convolutions. Technically, it computes a factorization $H = CGA$ of the convolution matrix H (Eq. 2.4) [13], where G is a diagonal matrix. Given two lengths $M, N \in \mathbb{N}$, its parameters of the Toom-Cook algorithm are the $M + N - 1$ supporting points α_i. Ideally, the α_i are chosen to be small integers, like 0 and ± 1. Then the matrices C and A consist of $0, \pm 1$, making it possible to realize them by a number of pre- and post-additions, but without multiplications. For longer convolutions further supporting points α_i have to be chosen. This requires to use larger integers like $\pm 2, \pm 4, \ldots$, which show up within the matrices C and A. Unfortunately, this increases the number of multiplications significantly, making the Toom-Cook algorithm reasonable for short convolutions only.

2.4.2. Karatsuba algorithm

In 1962 Karatsuba and Ofman [54] published an algorithm for fast multiplication of large numbers with many digits, widely known as the *Karatsuba* algorithm. For several decades, the algorithm was the de facto fastest known multiplication algorithm for practical problem sizes. This made it a very important tool for the design of hardware multipliers in integrated circuits. Although the original Karatsuba algorithm was originally conceived for fast multiplication of integers, it can also be applied to other algebraic structures, allowing fast polynomial multiplication and as well fast convolution. Blahut [13] presented how the Karatsuba technique, which can be considered a special case of the modified Toom-Cook algorithm, can be applied for fast convolution. Hurchalla [47] showed, that the decomposing scheme of the Karatsuba algorithm conforms with real-time constraints, making the algorithm applicable for low latency filtering. The Karatsuba algorithm is a classic example for a divide-and-conquer strategy. Let $x(n)$ and $h(n)$ be two length-$2N$ sequences. Both sequences are split in half, forming four length-N subsequences

$$x_0(n) = x(0), \ldots, x(N-1) \qquad h_0(n) = h(0), \ldots, h(N-1)$$
$$x_1(n) = x(N), \ldots, x(2N-1) \qquad h_1(n) = h(N), \ldots, h(2N-1)$$

The $2N \times 2N$-linear convolution $x(n) * h(n)$ can be expressed using four $N \times N$-sub convolutions [47]

$$y_0(n) = x_0(n) * h_0(n) \tag{2.20}$$
$$y_1(n) = x_0(n) * h_1(n) + x_1(n) * h_0(n) \tag{2.21}$$
$$y_2(n) = x_1(n) * h_1(n) \tag{2.22}$$

The desired output sequence $y(n)$ of length $4N - 1$ is overlap-added from the sequences $y_0(n), y_1(n), y_2(n)$ of length $2N - 1$ (as illustrated in Fig. 2.4)

$$y(n) = y_0(n) + y_1(n - N) + y_2(n - 2N) \tag{2.23}$$

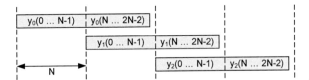

Figure 2.4.: Overlap-Add scheme used in the
Karatsuba convolution algorithm

The Karatsuba algorithm founds on the following equivalence

$$
\begin{aligned}
y_1(n) &= x_0(n) * h_1(n) + x_1(n) * h_0(n) \\
&= [\,x_0(n) + x_1(n)\,] * [\,h_0(n) + h_1(n)\,] \\
&\quad - \underbrace{x_0(n) * h_0(n)}_{=y_0(n)} - \underbrace{x_1(n) * h_1(n)}_{=y_2(n)}
\end{aligned}
\tag{2.24}
$$

It allows computing the Eq. 2.20-2.22 using three $N \times N$-sub convolutions instead of four, thus saving one $N \times N$-sub convolution. However, this comes at the expense of $2N$ further additions $([x_0(n) + x_1(n)], [h_0(n) + h_1(n)])$ and $2 \times (2N - 1)$ subtractions $(-y_0(n), -y_2(n))$. The full potential of this trick is unleashed, when the equivalance is applied recursively. Therefore $N = 2^k (k \in \mathbb{N})$ is considered to be a power of two. This results in the Karatsuba convolution algorithm, where two length-N sequences (N a power of two) are recursively decomposed and the reduction Eq. 2.24 is applied in every stage. The decomposition ends in trivial 1×1-convolutions $y(0) = x(0) \cdot h(0)$.

Computational complexity

The runtime complexity of the Karatsuba multiplication algorithm is $\mathcal{O}(N^{\log_2 3})$ [57]. This class also holds for the Karatsuba convolution method. The actual number of operations depends on the implementation. A detailed runtime analysis of the presented algorithm is carried out in the following, revealing the exact number of arithmetic operations, which is used for later comparisons.

The trivial case of an 1×1-convolution requires a single multiplication ($y_0 = x_0 \cdot h_0$). Let now $N = 2^k (k \in \mathbb{N})$ be a power of two. Each $N \times N$-Karatsuba convolution recursively computes three $N/2 \times N/2$-Karatsuba convolutions (Eq. 2.20, 2.22 and 2.24). Before this step, the intermediate sequences $[x_0(n) + x_1(n)], [h_0(n) + h_1(n)]$ have to be added, requiring $2 \times N/2 = N$ additions. Then, the subtractions in Eq. 2.24 must be performed on $2 \times (N - 1)$ overlapping elements. Finally, the computed intermediate sequence $y_1(n)$ is overlap-added to the (non-overlapping) sequences

$y_0(n)$ and $y_2(n)$. On close examination, this step consumes $N-2$ additions, as only $N-2$ samples actually overlap (see figure 2.4). Alltogether the number of arithmetic operations $T(N)$ is given by the recurrence

$$T(1) = 1, \qquad T(N) = 3T\left(\frac{N}{2}\right) + 4N - 4 \qquad (2.25)$$

It is helpful to rewrite this expression using the exponent k

$$T'(0) = 1, \qquad T'(k) = 3T'(k-1) + 4 \cdot 2^k - 4 \qquad (2.26)$$

By nesting, this equation can then be transformed into the explicit form

$$T'(k) = 7 \cdot 3^k - 8 \cdot 2^k + 2 \qquad (2.27)$$

Reinserting the definition of N finally yields the total number of arithmetic operations for a $N \times N$-Karatsuba convolution

$$T(N) = 7 \cdot 3^{\log_2 N} - 8N + 2 \qquad (2.28)$$

The runtime complexity class of the algorithm is found by applying the *Master theorem* [25] to Eq. 2.26. This yields the known [57] runtime complexity class of $\mathcal{O}(N^{\log_2 3})$. Considering that $\mathcal{O}(N^{\log_2 3}) \subset \mathcal{O}(N^2)$ the Karatsuba algorithm can be considered a *fast* convolution algorithm, working entirely in the time-domain. Nevertheless, it is asymptotically slower than $\mathcal{O}(N \log N)$ convolution methods (e.g. FFT convolution, cp. Sec. 2.5.2).

2.4.3. Improved Karatsuba convolution

A direct 2×2 linear convolution (Eq. 2.3) has the form

$$[\, x(0) \cdot h(0), \quad x(0) \cdot h(1) + x(1) \cdot h(0), \quad x(1)) \cdot h(1) \,] \qquad (2.29)$$

and requires five arithmetic operations (four multiplies and one addition). In contrast, a 2×2 Karatsuba convolution is slightly more complex

$$[\, \underbrace{x(0) \cdot h(0)}_{=y_0(0)}, \quad (x(0) + x(1)) \cdot (h(0) + h(1)) - y_0(0) - y_2(0), \quad \underbrace{x(1) \cdot h(1)}_{=y_2(0)} \,] \qquad (2.30)$$

consuming seven operations (three multiplies, two additions and two subtractions). Hence, the Karatsuba 'trick' turns out not to be beneficial in every case. As the method decomposes a large convolution into a large number of small 2×2 convolutions, these small convolutions do strongly affect the overall computational complexity of the entire method. Therefore, the author suggests two improvements over the original method [47]: (1) By implementing 2×2 convolutions using direct convolution, the performance can be significantly improved. Table 2.1 shows the number of arithmetic operations for $N \times N$ linear convolutions, including the direct approach and the Karatsuba variants. Apart from interpolation-based approaches, algorithms for very short lengths ($N < 10$) can be derived by hand, offering some additional savings. Several authors list such short convolution templates—for instance for linear convolution [5, 47] and for circular convolution [73, 13]. Agarwal and Cooley propose a 4×4 linear convolution template that requires 20 operations (five multiplies and 15 additions), saving five operations over the direct convolution. The proposed method (2) combines it with the Karatsuba algorithm, so that the Karatsuba decomposition stops at $N = 4$ and then the template is used. The second approach saves even more operations. In order to constrast the results, table 2.1 also includes the complexity of FFT-based convolution [105] (see Sec. 2.5.2). The data in table 2.1 is visualized in figure 2.5. The numbers consider an implementation using three $2N$ real-data FFTs [50]. It can be seen that the Karatsuba algorithm is not only fast by means of its asymptotic behaviour, but also in the actual number of operations. This is remarkable in face of that the algorithm works entirely in the time-domain. The improved variant (2) outperforms the direct approach for length $N \geq 4$. Interestingly, it also outperforms the FFT-based convolution for lengths $N < 128$. However, these break-even points might differ in practice. It turned out to be rather challenging to fully develop the Karatsuba technique with respect to exploitation of the available processor capabilities. In contrast high-performance FFT libraries are technically very matured (cp. FFT algorithms in Sec. 2.5.2 and benchmarks in Sec. 7.4).

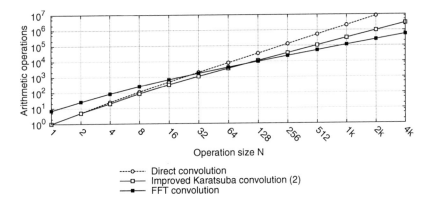

Operation size N

---o--- Direct convolution
—□— Improved Karatsuba convolution (2)
—■— FFT convolution

Figure 2.5.: Arithmetic complexity of direct $N \times N$ linear convolution, Karatsuba convolution and real-data FFT-based fast convolution (note, that both axis have logarithmic scales)

Length N	Direct convolution (Eq. 2.11)	Karatsuba convolution (Eq. 2.28)	Karatsuba convolution improv. (1)	Karatsuba convolution improv. (2)	FFT-based convolution (Eq. 2.48)
1	1	1	–	–	7
2	5	8	5	–	26
4	25	33	27	20	86
8	113	127	109	88	254
16	481	441	387	324	686
32	1,985	1,447	1,285	1,096	1,730
64	8,065	4,593	4,107	3,540	4,190
128	32,513	14,287	12,829	11,128	9,830
256	130,561	43,881	39,507	34,404	22,574
512	523,265	133,687	120,565	105,256	50,954
1,024	2,095,105	405,153	365,787	319,860	113,534
2,048	8,384,513	1,223,647	1,105,549	967,768	250,286
4,096	33,546,241	3,687,321	3,333,027	2,919,684	547,022

Table 2.1.: Arithmetic complexity of direct $N \times N$ linear convolution, Karatsuba convolution and real-data FFT-based fast convolution

2.4.4. Conclusions

The Karatsuba algorithm is an important case of the general class of Toom-Cook algorithms. It is applicable only for power-of-two sizes $N = 2^k$. However, Toom-Cook algorithms can be derived for many other combinations of lengths $M \times N$ [13]. Hurchalla [47] presents decompositions into three, four and five parts.

From the results in table 2.1 it can be concluded, that interpolation-based fast convolution algorithms offer significant computational savings over direct time-domain filtering. Here, the author showed how the Karatsuba technique can be further improved by the use of optimized convolution templates. For short lengths, these methods might also be an alternative to FFT-based convolution—given a well-thought implementation.

However, interpolation-based convolution methods can also be seen from a different perspective: they provide efficient schemes to decompose large convolutions into small ones. The fundamental constraint for real-time processing is that these schemes are real-time capable (detailed discussion of these constraints in chapter 6). Hurchulla [47] showed for the Karatsuba algorithm, that even with full nesting, all partial convolution results can be computed in time, making the algorithm real-time compliant. The algorithm does not introduce additional latencies, as all of its nested sub convolutions can be computed in time. However, there is no general 'zero-latency' property for the theoretical framework of interpolation-based convolution. Examples of Toom-Cook algorithms can be found [47], where the execution order of the sub convolutions inherits timing dependencies, that result in a large latency.

2.5. Transform-based fast convolution

The most important class of fast convolution algorithms, with most frequent use in practice, are *transform-based* fast convolution methods. They found on two cornerstones: Firstly, a discrete transform, in whose domain the convolution operation is simpler to compute (lower complexity class). Secondly, a *fast* algorithm to compute the transform in less than $\mathcal{O}(N^2)$, typically $\mathcal{O}(N \log N)$.

A method that revolutionized fast convolution is FFT-based fast convolution, published by Stockham [105] in 1966, one year after Cooleys and Tukeys renowned FFT algorithm [24]. Today it is still the most widely known and commonly used fast convolution technique and became widely a synonym for *fast convolution*. The cyclic convolution property (CCP) of the DFT states, that circular convolution in the time-domain corresponds to the pair-wise multiplication of spectral coefficients in the DFT domain. The mathematical origin of this relation can be traced back to matrix diagonalization in linear algebra. By computing DFTs using FFT algorithms, a fast circular convolution method in $\mathcal{O}(N \log N) \subset \mathcal{O}(N^2)$ is obtained. With appropriate zero-padding, the technique can be used to realize fast linear convolution as well.

The DFT demands complex-valued arithmetic, which especially on those early day computers was rather expensive to realize. This motivated research in real-valued transforms. A promising alternative to the DFT was the discrete Hartley transform (DHT). Other considered candidates are discrete trigonometric transforms (DTTs), including the variety of discrete sine transforms (DSTs) and discrete cosine transforms (DCTs). Fast algorithms exists for all of the cited transforms. Unfortunately, none of them holds the strict CCP and thus, spectral convolutions become more than just a pairwise product of spectral coefficients. For discrete trigonometric transforms (DTTs) in particular, such a convolution multiplication property (CMP) only exists for special convolution operators. Yet the real-valued transform techniques are worth the consideration as an alternative to FFT-based convolution, in particular when hardware transformers are available or the input data is already within the domain of the transforms.

All of the above mentioned transforms are defined over trigonometric functions and hence rely on *floating-point arithmetic*. Interestingly, integer-based transforms with similar properties as the DFT exist. A starting point was the extension of the DFT to finite fields by Pollard [75] in the 1970s. Instead of using the complex number plane, a transform of similar structure to that of the DFT can be defined in modulo arithmetic over integers. In successive years a variety of methods has been proposed, which constitutes the class of number theoretic transforms (NTTs). These transforms have the CCP. The existence of fast NTTs algorithms allows realizing a numerically exact fast convolution without round-off errors. Especially for the design of hardware convolution units, they turned out to be very beneficial, reducing the complexity of circuits. However, number theoretic transform (NTT)-based convolution never reached the same level of popularity as FFT-based convolution.

This section reviews transform-based convolution techniques in the face of real-time filtering and identifies the computationally most efficient one. Aspects of hardware design are not considered here. The focus lies on software implementations on current general purpose processors. A main part of the section is dedicated to the review of FFT-based convolution. Improvements to the original FFT-based convolution technique are discussed. The computational complexity of the FFT convolution techniques is carefully examined. Afterwards, real-valued transforms and NTTs are discussed as alternative techniques. It is outlined how convolution can be realized in the domains of these transforms. The complexity is analyzed and is compared to the one of the FFT-based approach, determining if they are actual alternatives.

2.5.1. Discrete Fourier Transform

The length-N discrete Fourier transform (DFT) [73, 21, 74] transforms a sequence $x(n) = x(0), \ldots, x(N-1)$ into a N-point DFT spectrum $X(k) = X(0), \ldots, X(N-1)$ (Eq. 2.31). $x(n)$ can be recovered from the DFT spectrum $X(k)$ using an inverse DFT (Eq. 2.32).

$$x(n) \overset{\mathcal{DFT}}{\circ\!\!-\!\!\bullet} X(k) = \sum_{n=0}^{N-1} x(n) e^{-2\pi i \frac{nk}{N}} = \sum_{n=0}^{N-1} x(n) W_N^{nk} \qquad (2.31)$$

$$X(k) \overset{\mathcal{DFT}^{-1}}{\bullet\!\!-\!\!\circ} x(n) = \frac{1}{N} \sum_{k=0}^{N-1} X(k) e^{2\pi i \frac{nk}{N}} = \frac{1}{N} \sum_{k=0}^{N-1} X(k) W_N^{-nk} \qquad (2.32)$$

The kernels W_N^{nk} and W_N^{-nk} of the DFT and IDFT are powers of the Nth primitive root of unity $W^{-2\pi i/N}$ in the complex number plane $\mathbb{C}$. In general, the input and output values are complex numbers $x(n), X(k) \in \mathbb{C}$. Eq. 2.31 and 2.32 can also be interpreted as matrix-vector products of the form $\vec{X} = F_N \vec{x}$ (DFT) and $\vec{x} = F_N^{-1} \vec{X}$ (IDFT), with vectors $\vec{x}, \vec{X} \in \mathbb{C}^N$ and DFT matrices $F_N, F_N^{-1} \in \mathbb{C}^{N \times N}$.

$$F_N = \left(W_N^{jk} \right)_{0 \le j,k < N} \qquad F_N^{-1} = \frac{1}{N} \left(W_N^{-jk} \right)_{0 \le j,k < N} \qquad (2.33)$$

Circular convolution property

Within the DFT domain, the circular convolution $\tilde{y}(n) = x(n) \circledast h(n)$ (Eq. 2.12) of two length-N sequences $x(n)$ and $h(n)$ can then be realized by the element-wise multiplication of DFT coefficients (symbolized by $\times$) [73, 21, 74]

$$\tilde{y}(n) = \mathcal{DFT}_{(N)}^{-1} \left\{ \mathcal{DFT}_{(N)}\{ x(n) \} \times \mathcal{DFT}_{(N)}\{ h(n) \} \right\} \qquad (2.34)$$

This relation is widely known and often referred to as the *cyclic convolution property* (*CCP*) [3, 5]. It can be derived by inserting the definition of circular convolution 2.12 into Eq. 2.31 (for details see [74]). Another way of interpretation arises from linear algebra. Written as products of matrices and vectors, Eq. 2.34 reads

$$\tilde{y}(n) = x(n) \circledast h(n) \;\; \cong \;\; \vec{\tilde{y}} = C\vec{x} \;\; \Leftrightarrow \;\; F_N^{-1} \underbrace{\left(F_N C F_N^{-1} \right)}_{=\text{diag}(H)} (F_N \vec{x}) \qquad (2.35)$$

The DFT matrix F_N (Eq. 2.33) is a Vandermode matrix consisting of powers W_N^{jk} of Nth primitive roots of unity. Its column vectors form the basis functions of the DFT, which are all orthogonal. Hence, F_N has rank$(F_N) = N$ and defines a base of the vector space $\mathbb{C}^N$. Applying the DFT in form of a right-multiplication $F_N \vec{x}$ can therefore be interpreted as a change of basis (time-domain $\rightarrow$ DFT domain). The right-product $F_N^{-1} \vec{X}$ corresponds to the inverse transformation (DFT domain $\rightarrow$ time-domain). Expressing a circulant matrix C (Eq. 2.13) within this 'discrete Fourier basis' turns it into a diagonal matrix

$$F_N C F_N^{-1} = \text{diag}(H(0), \dots, H(N-1)) \qquad (2.36)$$

$$\text{with} \quad H(k) = \mathcal{DFT}_{(N)}\{\, h(n)\,\}$$

This holds because the DFT basis (Eq. 2.33) defines an eigenvector basis of the circulant matrix (cp. Eq. 2.15). The elements of the diagonal matrix are the discrete Fourier coefficients $H(k)$ of the sequence $h(n)$. These correspondences are illustrated in example 2.2 for the case of $N = 4$.

2.5.2. Fast Fourier Transform

Shortly after Cooley and Tukeys important FFT paper [24] in 1965, Stockham proposed to use the FFT for fast convolution and correlation [105]. Until today FFT-based convolution is arguably the most applied fast convolution concept. This section firstly reconsiders the basic (unpartitioned) method as suggested by Stockham [105]. Its performance is largely determined by the FFT algorithm, which accounts for the major share of the runtime. Hence, significant improvements can be achieved by exploiting specific properties of the DFT and its computation using FFT techniques. Therefore, the most important FFT algorithms, their historic development and incorporated concepts are briefly reviewed subsequently. Afterwards, improvements to the basic FFT-convolution technique are discussed, in particular those which are of interest for audio processing (e.g. real input data). The study includes only those aspects which are relevant for comparing the basic FFT convolution technique to other fast convolution methods. A more detailed discussion of improvements is given in chapter 4, also with respect to computing running convolution, which is not part of this chapter. In order to analyze the complexity bounds, special attention is payed to partial transforms. These can be applied to lessen the increased computation that comes along with the necessary zero-padding in order to achieve linear convolution. Concepts of partitioning are not reviewed in this section. They are regarded in the chapters 3-6.

$x(n) = \begin{bmatrix} 2 & -1 & 3 & 1 \end{bmatrix}$

$h(n) = \begin{bmatrix} 1 & 2 & 3 & -1 \end{bmatrix}$

$\vec{x} = \begin{bmatrix} 2 & -1 & 3 & 1 \end{bmatrix}^{\mathsf{T}}$

$\vec{h} = \begin{bmatrix} 1 & 2 & 3 & -1 \end{bmatrix}^{\mathsf{T}}$

$\tilde{y}(n) = x(n) \circledast h(n)$

$\quad = \begin{bmatrix} 14 & 3 & 6 & 2 \end{bmatrix}$

$\vec{\tilde{y}} = C \cdot \vec{x}$

$$= \underbrace{\begin{bmatrix} 1 & -1 & 3 & 2 \\ 2 & 1 & -1 & 3 \\ 3 & 2 & 1 & -1 \\ -1 & 3 & 2 & 1 \end{bmatrix}}_{=C} \cdot \begin{bmatrix} 2 \\ -1 \\ 3 \\ 1 \end{bmatrix}$$

$= \begin{bmatrix} 14 & 3 & 6 & 2 \end{bmatrix}^{\mathsf{T}}$

$\mathcal{DFT}_4$

$$F_4 = \begin{bmatrix} 1 & 1 & 1 & 1 \\ 1 & -i & -1 & i \\ 1 & -1 & 1 & -1 \\ 1 & i & -1 & -i \end{bmatrix}$$

$\mathcal{DFT}_4^{-1}$

$$F_4^{-1} = \frac{1}{4} \begin{bmatrix} 1 & 1 & 1 & 1 \\ 1 & i & -1 & -i \\ 1 & -1 & 1 & -1 \\ 1 & -i & -1 & i \end{bmatrix}$$

$X(k) = \mathcal{DFT}_4\{ x(n) \}$

$\quad = \begin{bmatrix} 5 & -1+2i & 5 & -1-2i \end{bmatrix}$

$\vec{X} = F_4 \cdot \vec{x}$

$\quad = \begin{bmatrix} 5 & -1+2i & 5 & -1-2i \end{bmatrix}^{\mathsf{T}}$

$H(k) = \mathcal{DFT}_4\{ h(n) \}$

$\quad = \begin{bmatrix} 5 & -2-3i & 3 & -2+3i \end{bmatrix}$

$F_4 \cdot C \cdot F_4^{-1} =$

$$\begin{bmatrix} 5 & 0 & 0 & 0 \\ 0 & -2-3i & 0 & 0 \\ 0 & 0 & 3 & 0 \\ 0 & 0 & 0 & -2+3i \end{bmatrix}$$

$Y(k) = X(k) \cdot H(k)$

$\quad = \begin{bmatrix} 25 & 8-i & 15 & 8+i \end{bmatrix}$

$\vec{Y} = \mathrm{diag}(5, -2-3i, 3, -2+3i) \cdot \vec{X}$

$\quad = \begin{bmatrix} 25 & 8-i & 15 & 8+i \end{bmatrix}^{\mathsf{T}}$

$\tilde{y}(k) = \mathcal{DFT}_4^{-1}\{ Y(k) \}$

$\quad = \begin{bmatrix} 14 & 3 & 6 & 2 \end{bmatrix}$

$\vec{\tilde{y}} = F_4^{-1} \cdot \vec{Y}$

$\quad = \begin{bmatrix} 14 & 3 & 6 & 2 \end{bmatrix}^{\mathsf{T}}$

Example 2.2: DFT-based length-4 circular convolution and corresponding matrix-vector representations (C circulant matrix, F_4 4-point DFT matrix, F_4^{-1} 4-point inverse DFT matrix)

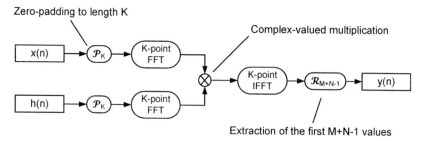

Figure 2.6.: Fast linear convolution realized by circular convolution in the DFT domain, implemented using fast Fourier transforms (FFTs), as proposed by Stockham [105]

FFT-based fast convolution

Let $x(n)$ be a length-M and $h(n)$ a length-N sequence. The $M \times N$ linear convolution $x(n) * h(n)$ in $\mathcal{O}(K \log K)$ can be computed in $\mathcal{O}(K \log K)$, by implementing Eq. 2.34 using K-point FFTs. The transform size K must satisfy $K \geq M + N - 1$ (Cond. 2.18), otherwise the linear convolution is corrupted (see section 2.5.1). Figure 2.6 illustrates this well-known method, which was proposed by Stockham [105].

Initially both sequences are padded to length K, expressed by a padding operator $\mathcal{P}_K$ and transformed using K-point forward FFTs. The resulting K-point DFT spectra $X(k)$ and $H(k)$ are pairwisely multiplied using K complex-valued multiplications. These are referred to as '*spectral convolutions*'. The result is transformed back into the time-domain using a K-point inverse fast Fourier transform (IFFT). Given that $K > M + N - 1$ the first $M + N - 1$ samples are selected, represented by the trim operator $\mathcal{R}_{M+N-1}$. The runtime complexity is a function of the transform size K, determined by the sequence length M, N, of the following general form

$$T_{\text{CONV-FFT}}(K) := 2\,T_{\text{FFT}}(K) + T_{\text{CMUL}}(K) + T_{\text{IFFT}}(K) \qquad (2.37)$$

$$\text{with} \quad K \geq M + N - 1 \qquad (2.38)$$

The terms $T_{\text{FFT}}(K)$, $T_{\text{IFFT}}(K)$ and $T_{\text{CMUL}}(K)$ express the computational complexity (number of arithmetic operations, central processing unit (CPU) cycles or runtimes). The general formula in Eq. 2.37 will be refined in later sections and chapters. Possible points for optimization can already be identified here: transformations and spectral convolutions. The choice of the transform size K is crucial for the performance of the algorithm. This aspect is reviewed in detail in section 4.1.4. If the filter $h(n)$ remains unchanged, the spectrum $H(k)$ can be reused for multiple input signals and one forward

FFT can be dropped from Eq. 2.37. By time-reversing one sequence the method can be used to compute fast correlations [105]. General DFT spectra consist of complex-valued coefficients. Real-valued input data introduces symmetries to the DFT spectra, so that effectively only half of the coefficients need to be stored and processed. The details are examined later, at the end of this section. The inevitable zero-padding of the input sequences is unfortunately carried into the computation of the FFTs. Transform algorithms process many ineffective zeros. Techniques to save operations from these circumstances are reviewed in section 2.5.4 on partial transforms. A further simplification arises for even or odd sequences, which have purely real respectively imaginary DFT coefficients. The resulting arithmetic simplifications are examined in section 2.5.3.

Computing the Fast Fourier Transform

This section gives a brief overview on the current state of the art in computing the fast Fourier transform. The focus lies on the algorithmic design principles and in particular on the decomposition techniques for obtaining divide-and-conquer algorithms for the fast computation of the DFT. Only the most important milestones with respect to fast convolution are outlined here, serving as the basis for the discussion of enhancements in FFT-based fast convolution techniques. These do not regard the FFT as a black box, but try to exploit certain properties of the transform and the transform algorithms. A compact overview on the state of the art in computing the FFT can be found in [29]. For a more comprehensive introduction, including the mathematical backgrounds, the author recommends the textbooks by Nussbaumer [73] and Burrus and Parks [21].

History of the FFT

The FFT is presumably one of the most intensively researched algorithms in history and numerous papers have been published about it [115]. In the year 2000 issue of 'computing in science and engineering' the FFT was credited one of the 'top ten algorithms of the century' [26] with the 'greatest influence on the development and practice of science and engineering in the 20th century'. In 1965 John W. Tukey and James W. Cooley published a groundbreaking paper on the fast computation of discrete trigonometric series [24]. This publication is widely considered as 'the original fast Fourier transform'. In their paper they derive the famous radix-2 DIT FFT algorithm as an *example*, which lead to the widespread belief, that the Cooley-Tukey FFT is only suitable to compute power-of-two FFTs. But it is not. The scientific impact of Cooley's and Tukey's algorithm is reasoned by the fact, that they developed an algorithm for *any* FFT length N which is not a prime number. The essential contribution of their paper is the composition of a large DFT into several DFTs of smaller sizes by utilizing an index-mapping technique called common-factor map (CFM). The use of this mapping requires intermediate multiplication by complex constants, called *twiddle factors* [40], which account for the most computation.

A thorough study in the mid 1980s [45] revealed that the German mathematician Carl Friedrich Gauss can be considered the first inventor of the FFT. Gauss developed an algorithm which is mathematically equivalent to the Cooley-Tukey twiddle-factor FFT. The reader is encouraged to read the interesting paper by Heideman, Johnson and Burrus [45], from which some key findings are cited in the following. Gauss who used his algorithm for computing the trajectories of asteroids, did not publish the algorithm in his lifetime. It was rediscovered posthumously from his collected publications. His invention cannot be dated accurately, but very likely it even predates Fourier's publications on harmonic analysis in 1807 [45]—the origin of the discrete Fourier transform. In the following 150 years several *fast* algorithms for discrete trigonometric series have been invented, published and used [45] (first for some fixed length $4, 8, 12, 16, 32$, later for multiples of powers $2^n k, 3^n k$). However, none of these results were as general as Gauss' and Cooley's and Tukey's results and could be applied for arbitrary composite transform sizes. A milestone before the Cooley-Tukey FFT was the Good-Thomas prime-factor algorithm (PFA) [42, 109], published 1958. It allows the efficient computation of length-N DFTs, when N is assembled from *coprime* factors (e.g. $15 = 3 \cdot 5$). Thereby it makes use of an index mapping called prime-factor map (PFM). It cannot be used for factorizations with common factors (e.g. $20 = 2 \cdot 10$ where $\gcd(2, 10) = 2$). The prime-factor algorithm does not require multiplications with intermediate twiddle factors, like for the Cooley-Tukey FFT, but unfortunately it does involve the computation of many prime-size DFTs. Today the PFA still serves as a decomposition principle for computing FFT of arbitrary sizes, that are not primes. Further remarkable achievements were the publication by Rader [77] in 1968 and Bluestein [14] in 1970, who showed how DFTs can be computed by circular convolutions. In particular Raders method reexpresses an N-point prime-size DFT as an $N-1$-point circular convolution, with the help of the discrete logarithm [77]. Eventually, this circular convolution is then realized using regular FFT techniques again, which do not involve prime-size transforms anymore. This approach paved the way to fast $\mathcal{O}(N \log N)$ algorithms for computing prime-size FFTs. Yet it should be remarked, that these prime-size FFTs compute several magnitudes slower than power-of-two FFTs (see benchmarks in chapter 7). Winograd [127] suggested to use Rader's prime-size FFTs concept in conjunction with the PFA, where it is used to compute prime-size FFT efficiently. This algorithm, known as the Winograd Fourier transform algorithm (WFTA), also provides a lower bound for the number of multiplications for an FFT. Unfortunately, the algorithm comes along with an increased number of additions, making other techniques faster in practice. Based on the original Cooley-Tukey CFM mapping, several other radices (radix-4, radix-8, etc.) have been developed and it has been shown that these transforms can save further operations over the classical radix-2 FFT [10]. A major step forward in reducing the arithmetic complexity of the FFT happened with the invention of the *split-radix FFT*. Originally conceived by Yavne in 1968 [129] it remained largely unnoticed for many years [115]. Split-radix FFTs decompose an N-point DFT into two $N/4$-point DFTs and one $N/2$-point DFT. Duhamel and Holland [27] reinventend this decomposition concept in 1984 and popularized it. A split-radix FFT can be computed for any transform length N that is dividable by 4. Until today split-radix FFTs achieve the lowest operation counts for power-of-two sizes N [115, 27, 50]. At the time of writing, the most recent and advanced FFT algorithms for complex and real data for power-of-two sizes were the split-radix algorithms by Johnson and Frigo [50]. Not only these algorithms are part of the *FFTW* library [34], but also do the authors provide exact formulas for the number of arithmetic operations. In this thesis, these two algorithms are mostly used for theoretical comparisons.

Besides this line of developments aiming for general FFT transforms, a scientific topic of great interest became specialized transforms for real-valued input data, known as *real-data FFTs*. By exploiting the complex-conjugate symmetry in DFT spectra of real-valued input data, these algorithms compute an N-point real-data FFT in approximately 50-60% of the runtime of a complex-valued N-point transform. The application of the FFT for digital filtering was proposed by Stockham in 1966, shortly after the reinvention of the FFT by Cooley and Tukey. Stockham [105] showed how the FFT could be used for fast convolution and fast correlation. Realizing fast linear convolution using fast circular convolution in the DFT domain, involved zero-padding of the input sequences. When computing running convolutions using Overlap-Save (OLS), a subset of the IDFT output values is needed. For these cases partial FFT transforms have been proposed. These concepts will be reviewed in detail in section 2.5.4. The computational savings by partial transforms for digital filtering turn out to be little. Many of the algorithmic concepts derived for FFTs apply to other discrete transforms as well [29]. This includes the discrete Hartley transform, the family of discrete trigonometric transforms, as well the NTTs . For all of these transforms fast $\mathcal{O}(N \log N)$ algorithms exist.

Computing FFTs today

Nowadays, FFTs are mostly computed using matured high performance libraries, like $FFTW$ [34], Intel's *Math Kernel Library* (*MKL*) or *Performance Primitives* (*IPP*), AMD's *Math Core Library* (*ACML*) or Apples *vDSP* library. These libraries can be considered as collections of FFT building blocks. FFTW is thereby of particular scientific interest, as it is open-source software and its internal design is well-documented [36]. Furthermore, it computed the most complete set of discrete transforms of all mentioned libraries. The libary consists of a set of small-size fast Fourier transform (FFT) assembly language templates, called *codelets* (e.g. a length-32 real-data DFT). These templates can be specific for machines or instruction sets. On top of this, the libray holds various decomposition schemes (e.g. a radix-2 Cooley-Tukey decomposition). Before an FFT can be computed, the user has to *plan* the transform, providing the library the transform properties (real vs. complex, dimensions, data alignment, etc.). Then FFTW figures out the most performant decomposition of the desired transform maximizing the execution speed. This process can be guided and adjusted for other attributes (e.g. numerical stability). The library has been extensively benchmarked by the author for this thesis. The results are presented in chapter 7.

Real-valued input data

In audio processing, the input data and filter coefficients are usually real-valued. The N-point DFT (Eq. 2.31) is a complex-valued transform $\mathcal{DFT}_{(N)}$: $\mathbb{C}^N \rightarrow \mathbb{C}^N$, mapping N complex input values $x(n) \in \mathbb{C}$ to N complex DFT coefficients $X(k) \in \mathbb{C}$. For the case that all N input values are real $x(n) \in \mathbb{R}$, the corresponding DFT spectrum $X = \mathcal{DFT}_{(N)}\{ x(n) \}$ holds the Hermitian symmetry [74]

$$X(k) = \overline{X(N-k)} \tag{2.39}$$

Hence, the full N-point DFT spectrum can be reconstructed from C *complex-conjugate symmetric* (*CCS*) DFT coefficients, where

$$C = \left\lceil \frac{N+1}{2} \right\rceil = \begin{cases} \frac{N}{2} + 1, & N \text{ even} \\ \frac{N+1}{2}, & N \text{ odd} \end{cases} \tag{2.40}$$

Real-data FFT algorithms exploit these symmetries and compute roughly twice as fast as their complex-valued counterparts. An N-point real-to-complex (R2C) FFT maps N real-valued samples to C DFT coefficients (Eq. 2.40). An N-point complex-to-real (C2R) IFFT is the corresponding inverse transform.

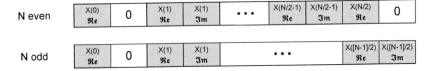

Figure 2.7.: Memory layout of complex-conjugate symmetric
N-point real-data DFT spectra

Since all $x(n) \in \mathbb{R}$, the first DFT coefficient $X(0) = \sum_{n=0}^{N-1} x(n)W_N^{n \cdot 0}$ (DC offset) is always real-valued, too. In case the transform size N is an even number, the central DFT coefficient $X(N/2)$ is purely real valued as well. Figure 2.7 shows the commonly used interleaved memory layout for complex-conjugate symmetric (CCS) DFT spectra. The zero imaginary parts can be dropped, resulting in the most compact representation, which is known as *packed format* [1]. With respect to symmetry, a real-data DFT effectively maps N floating point elements to N floating point elements (bijection). In the following the number of arithmetic operations required for spectra convolution of two real-valued DFT spectra is analyzed. A coarse estimate arises from the observation that only C complex-conjugate symmetric coefficients have to be multiplied, requiring $6C$ operations

$$T_{\text{CMUL-CCS}}(N) = 6C = \begin{cases} 6\left(\frac{N}{2} + 1\right) = 3N + 6 & N \text{ even} \\ 6\left(\frac{N+1}{2}\right) = 3N + 3, & N \text{ odd} \end{cases} \quad (2.41)$$

Looking into detail, all purely real-valued pairs of coefficients $X(0) \cdot H(0)$ and $X(N/2) \cdot H(N/2)$ for even N, can be multiplied using a single arithmetic operation, saving a couple of instructions

$$T_{\text{CMUL-CCS}}(N) = \begin{cases} 6\left(\frac{N}{2} - 1\right) + 2 = 3N - 4 & N \text{ even} \\ 6\left(\frac{N+1}{2} - 1\right) + 1 = 3N - 2, & N \text{ odd} \end{cases} \quad (2.42)$$

This precise statement complicates the theoretical analysis of algorithms, as conducted in the preceding chapters. The following simplification is often preferred

$$T_{\text{CMUL-CCS}}(N) \approx T_{\text{CMUL}}\left(\frac{N}{2}\right) = 3N \quad (2.43)$$

For reasonably large N this is a close estimate. For $N = 128$ the relative error is 1.05% (384/380) and $N = 1024$ it is 0.13% (3072/3068).

2.5.3. Symmetric filters

Sequences with an even or odd symmetry have DFT spectra which are either entirely real-valued or purely imaginary (Eq. 2.44 and 2.45) [74]. The general complex-valued the spectral convolutions simplify in these case, which allows saving operations. Input signals in audio processing are usually real-valued, but they do usually not inhere these symmetries. Their DFT coefficients are therefore complex-conjugate symmetric, but in general complex-valued. Neither the real nor the imaginary parts do completely vanish. In contrast, some types of filters hold the contemplated symmetries and allow further arithmetic simplifications. For symmetric respectively antisymmetric real-valued length-N impulse responses $h(n) \in \mathbb{R}$ the following relations hold

$$h(n) = h(N - n) \text{ even} \quad \overset{\mathcal{DFT}_{(N)}}{\circ\!\!-\!\!\bullet} \quad H(k) \text{ real} \quad (\Im\{H(k)\} = 0) \quad (2.44)$$

$$h(n) = -h(N - n) \text{ odd} \quad \overset{\mathcal{DFT}_{(N)}}{\circ\!\!-\!\!\bullet} \quad H(k) \text{ imag.} \quad (\Re\{H(k)\} = 0) \quad (2.45)$$

A class of filters that satisfies these properties are *causal generalized linear phase FIR filters* [74]. In acoustic virtual reality they are for instance used in the form of linear-phase wall-impedance filters in room acoustic simulation methods or as headphone equalization filters.

For the case that the impulse response is real and even, the spectral convolutions can be realized using just two multiplies (Eq. 2.46) instead of six operations per DFT coefficient. Accordingly, for an odd symmetry an additional change of sign is necessary (Eq. 2.47).

$$(a + bi)c = ac + bci \quad \text{(2 multiplies)} \quad (2.46)$$
$$(a + bi)di = -bd + adi \quad \text{(2 multiplies, 1 negation)} \quad (2.47)$$

Exploiting these symmetries in practice is hindered by the following constraints: the necessary zero padding (cp. Sec. 2.5.1) must be realized so that the symmetry properties (Eq. 2.44 or 2.45) are maintained. Otherwise, complex-valued DFT coefficients appear in $H(k)$. This can be assured by zero-padding on *both* sides and claiming that $h(0) = 0$. However, the savings in the arithmetic allow marginal speedups only, as the spectral convolutions demand little computation only (see figure 4.4). In partitioned convolution techniques the FFTs play a less dominant role and more computation falls back to spectral convolutions. Here however, the symmetry properties are sacrificed by the partitioning of the impulse response. Symmetries of a complete (unpartitioned) filter (symmetric or antisymmetric) do not translate into symmetries in the subfilters, thus making the simplifications unapplicable. In conclusion, only little computational savings can be expected for the case of symmetric filters.

Computational complexity

This thesis considers several fast convolution methods which are based on the FFT. As this chapter aims the comparison of general fast convolution concepts, one representative FFT-based convolution technique had to be selected. Chosen was the basic method in Fig. 2.6, implemented with real-valued FFT transforms and making use of the previously discussed symmetries. With these considerations, the computational costs for FFT-based convolution in Eq. 2.37 are refined to the following function

$$T_{\text{CONV-FFT}}(K) := 2T_{\text{FFT-R2C}}(K) + T_{\text{CMUL-CCS}}(K) + T_{\text{IFFT-C2R}}(K) \quad (2.48)$$

As before, the transform size K must hold condition 2.37 in order to achieve linear convolution, as it is desired. An $N \times N$ linear convolution ($N = 2^k$) is computed with a transform size of $2N$ points. FFTs are computed using the real-data split-radix FFT by Johnson and Frigo [50]. It is assumed that the inverse FFT has the same number of operations. Spectral convolutions account for the number of operations in Eq. 2.42. Table 2.2 lists the number of arithmetic operations for the basic operations (FFTs, spectral multiplications) and the entire algorithm. The data is completed by the costs of a direct linear convolution in the time-domain (Eq. 2.3). It can be seen, that the major computational load falls back on the FFTs, while spectral convo-

Length N	Real-data split-radix FFT, IFFT (Eq. in [50])	Complex multiplies (Eq. 2.42)	FFT-based convolution (Eq. 2.48)	Direct convolution (Eq. 2.11)
1	2	1	7	1
2	6	8	26	5
4	22	20	86	25
8	70	44	254	113
16	198	92	686	481
32	514	188	1,730	1,985
64	1,270	380	4,190	8,065
128	3,022	764	9,830	32,513
256	7,014	1,532	22,574	130,561
512	15,962	3,068	50,954	523,265
1,024	35,798	6,140	113,534	2,095,105
2,048	79,334	12,284	250,286	8,384,513
4,096	174,150	24,572	547,022	33,546,241

Table 2.2.: Arithmetic complexity of $N \times N$ linear convolutions implemented using the FFT (Fig. 2.6) and by direct convolution (Eq. 2.3)

lutions make up for a minor part only. Until the length $N = 16$ the direct convolution requires less operations. Then the FFT approach takes the lead. Vast savings due to the lower complexity class $\mathcal{O}(N \log N) \subset \mathcal{O}(N^2)$ become apparent for larger N. For $N = 4096$ the FFT approach is > 60 times more efficient than the direct convolution. The improved Karatsuba convolution (variant 2) (section 2.4.2) outperforms the FFT-convolution for small sizes $N < 128$. But also for sufficiently large N, the FFT method clearly wins ($\mathcal{O}(N \log N) \subset \mathcal{O}(N^{\log_2 3})$). For $N = 4096$ the Karatsuba technique (2) is ≈ 5.3 times less efficient.

2.5.4. Partial DFTs

The majority of FFT algorithms is designed and optimized to compute complete transforms, including all input and output points. Particularly in convolution algorithms, it often occurs that only a subset of the input or output points of a transform are of interest. Input samples and filters are often zero-padded to match the transform size. By considering only the nonzero elements in the FFT algorithm, arithmetic operations could be saved over computing a complete transform. Computing a K-point DFT from only $L < K$ nonzero input samples, is known as a *partial input DFT*. In practice, an FFT is often computed of blocks of samples from which half are zero $L = N/2$. These cases are illustrated in Fig. 2.8. They occur frequently in techniques for computing running convolutions. A detailed study of these method is given in chapter 4. This is typically the case for the input data in Overlap-Add (OLA) running convolutions and the transformation of sub-filters in partitioned convolution methods. It is referred to as a *half-input FFT* in the following. The case when only a subset of output points from an inverse FFT is needed, is referred to as a *partial output IDFT*. This case often occurs in OLS running convolution algorithms, where part of the IDFT results are discarded. The case when one half of an IFFT is desired is called a *half-output IDFT* in this work. Specialized algorithms for fast partial transforms could thereby turn out to be beneficial for speeding up the convolution.

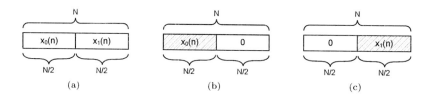

Figure 2.8.: Block decomposition of a signal into two halves

In this section, techniques for partial transforms are reviewed and discussed. The examination is based on partial FFTs, for which most concepts have been developed. The techniques reviewed here can be applied to several other transforms as well (e.g. NTTs), which share common mathematical properties (e.g. decomposition) and algorithmic origins (e.g. divide-and-conquer designs). Particular attention is paid to the benefit of these rather specialized algorithms in today's times of highly optimized, general purpose FFT libraries.

Goertzel algorithm

The Goertzel algorithm [41, 21, 13, 74] is a method allowing to compute individual coefficients of a DFT or IDFT, independent of each other. The algorithm is more efficient than the direct computation of a DFT coefficient $X(k)$ in Eq. 2.31. The starting point is to substitute part of the Fourier kernel W_N^j in Eq. 2.31 by the variable z. The summation of a single DFT coefficient $X(k)$ can be understood as evaluating the polynomial $p(z) = x(0)z^0 + \cdots + x(N-1)z^{N-1}$ in the point $p(z = W_N^k) = X(k)$. The principle behind the Goertzel algorithm is turning this computation into a recursive digital filter (see Fig. 2.9). Firstly, Horners rule [93] is applied to evaluate $p(z)$ efficiently. The resulting sequence of nested terms can be rewritten in form of a linear difference equation, defining a recursive filter. The samples $x(n)$ are given into the filter. After N filtering steps the output value $y(n)$ equals the desired DFT coefficient.

The straightforward derivation of the algorithm leads to the so-called first order Goertzel algorithm, which involves a single feedback path with a complex multiplication. This algorithm can be modified into a second-order Goertzel algorithm [74], as shown in figure 2.9. This structure is computationally more efficient by saving complex multiplications for real-valued input samples $x(n)$. The remaining complex multiplication with W_N^{-k} in the feed-forward branch only needs to be evaluated in the final N^{th} step [74] (marked gray).

The author analyzed the example code of second-order Goertzel algorithm in [21] and found the number of operations for computing a single coefficient $X(k)$ of an N-point DFT to be $6N + 8$. Consequently, the Goertzel algorithm's runtime to compute a complete N-point DFT is in $\mathcal{O}(N^2)$. This implies that the algorithm is only beneficial, when a very small number $L \ll N$ of DFT coefficients shall be computed. The break-even point, for that a complete N-point FFT is cheaper to compute than an L-point Goertzel algorithm can be roughly evaluated with $N/\log_2 N$ [74].

The author compared the second-order Goertzel algorithm's number of operations for real-valued inputs with a more recent real-data split-radix FFT algorithm [50]. Given a practical transform size of $N = 128$, the Goertzel algorithm is only more efficient when computing a *single* DFT coefficient ($L = 1$). Even for larger transforms $N = 4096$ the break-even point is $L \leq 3$. This renders the Goertzel algorithm unfeasible for partial transforms in fast convolution algorithms.

Pruned FFTs

A more efficient strategy for computing partial DFTs can be directly derived from fast Fourier transform algorithms. Markel [65] presented a technique based on the radix-2 decimation-in-frequency Cooley-Tukey FFT [24] for computing partial FFT, either with partial input data or partial output data. The computation on a FFT can be visualized using a flow graph (e.g. figure 2.10 and 2.11). The intermediate FFT computations are often referred to as *butterflies* [21], due to their crossed calculation paths. Computational savings are for a partial DFT achieved by *pruning* the flow graph and removing the computations of undesired values and computations that involve ineffective zeros. Further savings arise from the fact, that with pruning, several butterflies in an FFT flowgraph only need to be evaluated half, if one of the two input values is zero. For $L \ll N$ some butterflies in the flowgraph may be completely dropped, if both of its inputs are zero. Skinner [96] applied pruning to the DIT-FFT and improved Markel's method further, by reducing the number of twiddle factor multiplications in the first stages. Sreevinas and Rao [103] described a combined FFT algorithm with both, input and output pruning. The computational savings of a Markel pruned FFT over a complete FFT are approximately $n(l + 2(1 - 2^{-(n-l)}))^{-1} - 1$ [65], where $N = 2^n, L = 2^l$. The computational savings of both pruning concepts are analyzed in [100] in detail.

Markel's pruning [65] is illustrated in figure 2.10 in the example of an 8-point half-input FFT. The second half of input samples is considered to be zero

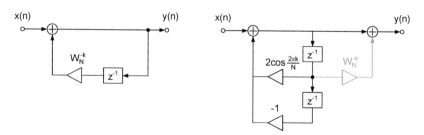

Figure 2.9.: First- and second-order Goertzel algorithms for computing individual DFT coefficients [74]

here $x(n \geq N/2) = 0, L = N/2$. Savings in the algorithm are achieved in the first butterfly, which only needs to be evaluated half. The remaining flowgraph remains identical. The computational savings over a complete FFT are 8 additions and 4 multiplications and thereby little. All 12 twiddle factor multiplications remain. Figure 2.11 shows the corresponding Skinner DIT-FFT. It saves as well 8 additions and 4 multiplications, but it requires only 8 twiddle factor multiplications. Both examples illustrate, that the computational savings for both algorithms are little. Further studies can be found in [100], where pruning is compared to modern split-radix FFTs. The authors conclude, that reasonable savings can only be achieved if a small number of points $L \ll N$ is considered. For $L \geq N/2$ they advise to compute full transforms instead.

A disadvantage of the pruning concepts is that they are essentially low-level techniques which modify an FFT algorithm. This does not go well along with current trends in FFT implementations and their extensive optimization. All present major FFT libraries (FFTW, IPP, MKL, ACML) are designed to compute full transforms and do not support pruning. Typically, a highly-optimized full transform computed with such a library will outperform a pruned FFT, that is derived by hand. The authors of FFTW [35] argue that the moderate savings by pruning are disproportionate to the effort of generating high-performance pruned implementations. They suggest not to consider pruning unless 1% or even less of the input/output values is needed [35]. A recent study [33] compares complete FFT to pruned FFT. Here, vectorized high-performance FFTs were specifically created using code generation techniques. This includes half-input and half-output pruned FFTs. The results of their practical benchmarks support the prior findings, that the savings for half-input or half-output FFT are little.

Transform decompositions

An elegant, systematic and even computationally more efficient alternative to pruning techniques are *transform decompositions* [100]. Based on a Cooley-Tukey decimation-in-frequency (DIF) decomposition [24], a large DFT is divided into several smaller DFTs using two-dimensional index mapping. The partial input data needs to be multiplied with twiddle factors. Then, regular complex FFTs are computed of these intermediate signals. The coefficients of the complete DFT are obtained by selection. In the following, the method by Sorensen and Burrus [100] is outlined.

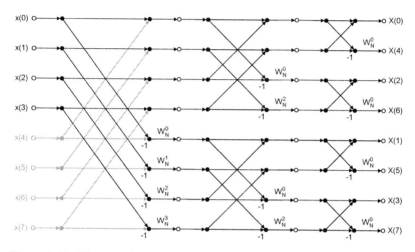

Figure 2.10.: Flow graph of an 8-point radix-2 decimation-in-frequency (DIF) input pruned[3] FFT [65]

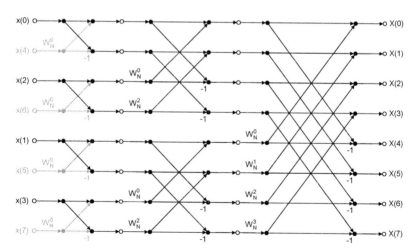

Figure 2.11.: Flow graph of an 8-point radix-2 decimation-in-time (DIT) input pruned[3] FFT [96]

---

[3]The second half of the input is zero $x(n \geq 4) = 0$

Transform decomposition requires a transform size N that can be factored into $N = P \cdot Q$ $(P, Q \in \mathbb{N})$. Then an index mapping of the following form can be defined

$$n = P \cdot n_1 + n_2 \qquad n_1 \in \{0, \ldots, Q-1\}, \quad n_2 \in \{0, \ldots, P-1\} \qquad (2.49)$$

$$k = k_1 + Q \cdot k_2 \qquad k_1 \in \{0, \ldots, Q-1\}, \quad k_2 \in \{0, \ldots, P-1\} \qquad (2.50)$$

Let $x(n)$ now be the length-N sequence of which an N-point DFT $X(k)$ shall be computed. Substituting the indices n, k in the definition of the DFT (Eq. 2.31) by the mapping in Eq. 2.50 yields [100]

$$X(k_1 + Qk_2) = \sum_{n_2=0}^{P-1} \sum_{n_1=0}^{Q-1} x(Pn_1 + n_2) W_N^{(Pn_1+n_2)(k_1+Qk_2)} \qquad (2.51)$$

$$= \sum_{n_2=0}^{P-1} \left[\sum_{n_1=0}^{Q-1} x(Pn_1 + n_2) W_N^{(Pn_1+n_2)k_1} \right] W_P^{n_2 k_2} \qquad (2.52)$$

The outer sum in Eq. 2.52 can be read as an P-point DFT

$$X_{k_1}(k_2) = \sum_{n_2=0}^{P-1} x_{k_1}(n_2) W_P^{n_2 k_2} \qquad (2.53)$$

of a signal $x_{k_1}(n_2)$, that is defined by the inner loop

$$x_{k_1}(n_2) = \sum_{n_1=0}^{Q-1} x(Pn_1 + n_2) W_N^{(Pn_1+n_2)k_1} \qquad (2.54)$$

A benefit is that Eq. 2.53 forms a DFT of a reduced transform size $P < N$. Note, that the inner sum does not form a Q-point DFT as it involves powers of the Nth root of unity W_N, instead of W_Q. Considering k_1 as a variable which runs from $0, \ldots, Q-1$ (Eq. 2.50), there are Q signals $x_{k_1}(n_2)$ and their P-point DFT transforms $X_{k_1}(k_2)$. From these the complete DFT spectrum $X(k)$ of the sequence $x(n)$ is obtained by simple selection

$$X(k_1 + Qk_2) = X_{k_1}(k_2) \qquad \Rightarrow \qquad X(k) = X_{k \bmod Q}\left(\left\lfloor \frac{k}{Q} \right\rfloor\right) \qquad (2.55)$$

The input signal $x(n)$ is accessed in a block-wise pattern $x(Pn_1 + n_2)$. It is partitioned into Q blocks of length P. When using the method for computing partial transforms, one chooses P, Q in a way, that only the first block ($n_1 = 0$) contains values $x(n_2) \neq 0$. In this case Eq. 2.54 simplifies to simple twiddle factor multiplications

$$x_{k_1}(n_2) = x(n_2) \cdot W_N^{n_2 k_1} \qquad (2.56)$$

Computing a partial input transform consists of three steps: Firstly, the P length-Q signals $x_{k_1}(n_2)$ are generated using (Eq. 2.56). Secondly, Q separate P-point DFTs are computed using regular FFTs, forming the spectra $X_{k_1}(k_2)$ (Eq. 2.53). In the third and last step $X(k)$ is recombined by selection according to Eq. 2.55.

Transform decompositions allow the computation with partial inputs and outputs. For $L \ll N$ the computational savings compared to a complete transform are immense. The technique clearly outperforms FFT pruning [100]. Another huge advantage over the pruning concepts is, that transform decompositions do not rely on modifications of the FFT algorithm. This is a strong argument in times where highly optimized FFT libraries are widely used. For a block-wise decomposition of the input $x(n)$, the modified sub signals in Eq. 2.54 will generally not vanish. This implies, that all Q FFTs of length P need to be computed. Given some transform size N, there are usually several possible combinations $N = P \cdot Q$. However, the authors [100] also point out that if $L \geq N/2$, it is still more beneficial to compute a complete split-radix FFT instead. These findings are consistent with prior observations for pruning.

Conclusions

Partial transforms seem not to offer computational benefits, unless a very limited number of input or output points is considered. Nevertheless, a potential application could be the unpartitioned convolution of short sequences with long sequences. Even if the transforms could be computed significantly faster, still an unnecessarily large number of DFT coefficients would have to be multiplied. Partitioned convolution methods (see chapter 3) overcome both of these disadvantages and reduce the zero-padding as well as the spectral convolutions to a minimum. In this context, half-input and half-output transforms (see figure 2.8) are very often computed. But the results indicate that if any computational savings can actually be achieved here, they are marginal.

The next sections are devoted to the question, if real-valued transforms offer computational benefits when realizing fast convolution with them.

2.5.5. Discrete Hartley Transform

An important transform in the history of digital signal processing is the discrete Hartley transform (DHT). It has been proposed by Bracewell [16] as the discretized version of the (continuous) Hartley transform. In the 1980s, the DHT was conceived as an alternative to the DFT, because it is purely real-valued and avoids complex-valued arithmetic. The DHT is closely related to the DFT. Due to this immediate relation, many of the algorithmic concepts for FFT apply for the DHT as well [101, 29] and similar fast Hartley transform (FHT) algorithms exist [17, 101, 28]. This section discusses the use of the FHT as an alternative to the FFT for implementing fast convolutions.

The discrete Hartley transform transforms a length-N real-valued sequence $x(n) = x(0), \ldots, x(N-1)$ into N real DHT coefficients, defined by [101]

$$x(n) \quad \overset{\mathcal{DHT}_{(N)}}{\circ\!\!-\!\!\bullet} \quad X(k) = \sum_{n=0}^{N-1} x(n) \operatorname{cas}\left(2\pi \frac{nk}{N}\right) \tag{2.57}$$

$$X(k) \quad \overset{\mathcal{DHT}^{-1}_{(N)}}{\bullet\!\!-\!\!\circ} \quad x(n) = \frac{1}{N} \sum_{k=0}^{N-1} X(k) \operatorname{cas}\left(2\pi \frac{nk}{N}\right) \tag{2.58}$$

'cas' denotes the *cosine-and-sine* function, also known as the Hartley kernel

$$\operatorname{cas}(x) = \cos(x) + \sin(x) = \sqrt{2} \cos\left(x - \frac{\pi}{4}\right) \tag{2.59}$$

It consists of cosines, shifted by $45°$ and scaled by the factor $\sqrt{2}$. The DHT is a linear operator and can be interpreted as matrix-vector product with an $N \times N$ transform matrix. Apart from a scale factor N^{-1}, the DHT is its own inverse $\mathcal{DHT}^{-1}_{(N)} = N^{-1}\mathcal{DHT}_{(N)}$. The Hartley kernel $\operatorname{cas}(2\pi \frac{nk}{N})$ does not form primitive N^{th} roots of unity in $\mathbb{R}$ and as a consequence of the results by Burrus [3], the DHT does not satisfy the strict cyclic convolution property (CCP) [28]. Nevertheless, circular convolution has an elegant and simple formulation in the DHT domain.

Circular convolution

For two length-N sequences $x(n), h(n)$ with DHT spectra $X(k) = \mathcal{DHT}_{(N)}\{x(n)\}$, $H(k) = \mathcal{DHT}_{(N)}\{h(n)\}$ the circular convolution $x(n) \circledast h(n)$ can be expressed in the DHT domain using the following relation [16, 28]

$$y = x \circledast h \quad \overset{\mathcal{DHT}}{\circ\!\!-\!\!\bullet} \quad Y = X(k) \cdot H_{\text{even}}(k) + X(-k) \cdot H_{\text{odd}}(k) \tag{2.60}$$

$H_{\text{even}}(k)$ and $H_{\text{odd}}(k)$ correspond to the even respectively odd part of the DHT spectrum $H(k)$. Inserting these definitions yields the expanded form

$$Y(k) = \frac{1}{2}\Big[X(k) \cdot (H(k)+H(N-k)) + X(N-k) \cdot (H(k)-H(N-k))\Big] \tag{2.61}$$

For the first DHT coefficient $Y(0)$ the Eq. 2.61 simplifies to $Y(0) = X(0) \cdot H(0)$. The paired additions $H(k) + H(N - k)$ and $H(k) - H(N - k)$ require $2N$ operations. Unless the filter h is modified, they can be precomputed once, saving additional operations. Then, the circular convolution in DHT domain consumes $1 + 3(N - 1) = 3N - 2$ operations. This equals the number of operations for circular convolution using the DFT for odd transform sizes N (Eq. 2.42). For even transform sizes N the DFT method consumes two fewer operations. Symmetric or anti-symmetric filters $h(n)$ cause the even part $H_{\text{even}}(k)$ respectively odd part $H_{\text{odd}}(k)$ to vanish, thus saving operations (cp. section 2.5.3).

Conclusions

The DHT formulation of circular convolution (Eq. 2.61) is not cheaper to realize than with real-data DFTs when symmetries are exploited (Eq. 2.42). The number of (real) multiplication operations is actually identical [28]. Unless the even and odd parts in Eq. 2.60 are given, the DHT technique is slightly more expensive, due to the necessary paired additions $H(k) + H(N - k), H(k) - H(N - k)$. Eq. 2.61 might be easier to vectorize than Eq. 2.34, as the purely real-valued pairwise multiplications are free from dependencies between real and imaginary parts as for complex multiplications. Nowadays, this argument is weakened by highly advanced instruction sets (e.g. SSE, AVX) which include mixed add-sub instructions as well as horizontal single-instruction multiple-data (SIMD) instructions. Moreover, for large DHT the paired additions do not hold the principle of data locality, which might influence the performance. It is questionable if the real-valued spectral convolutions have any computational benefit over the complex-valued in FFT-based convolution on current processors.

The strongest impact in performance has however the transform itself (most computation). It has been shown in the 1980s that FHT are marginally slower than real-data FFT algorithms [102]. Duhamel and Vetterli [28] suggest to prefer FFT-based convolution instead, also reasoned by the fact that the FFT has more general applications and has a greater importance. Maybe because of this, FHTs received less attention than FFT algorithms. The benchmarks on the test system (see section 7.4) revealed that fast Hartley transforms (FHTs) computed using FFTW are considerably slower than real-data FFTs. It is therefore concluded that FHT-based convolution does not form an alternative to FFT-based convolution nowadays.

2.5.6. Discrete Trigonometric Transforms

On the basis of cosine and sine kernels a variety of discrete transforms can be defined, which all have real-valued spectra coefficients [78]. This class of transforms became known as discrete trigonometric transforms (DTTs). Well-known examples are the different discrete cosine transforms (DCTs) and discrete sine transforms (DSTs). They slightly differ in their transform kernels with respect to the (assumed) symmetries of the input sequence. These differences matter in applications of the transforms. In particular the DCT-II is a fundamental tool in image and video compression (JPEG image coding and MPEG video coding standards). The DCT-IV plays an important role for lapped orthogonal transforms and serves as a basis for the modified discrete cosine transform (MDCT). The MDCT is widely used in audio compression techniques (MP3, Ogg Vorbis, Dolby Digital AC-3 and Advanced Audio Coding).

Discrete trigonometric transforms are closely related to the discrete Fourier transform (DFT). They can be derived from DFTs of symmetrically extended sequences [18]. Hence, one possibility to compute them is by using FFT algorithms [36, 81]. Fast native algorithms for DTT have been topic of intensive research in the last 30 years. Many papers on the topic exist in the literature–for software, but also direct hardware implementations. In the face of the extent, a formal definition of the multitude of DTTs is skipped here. An overview on the field can be found in the textbook by Britanak, Yip and Rao [18]. For all DTTs fast $\mathcal{O}(N \log N)$ algorithms exist. Much of the theory on computing fast Fourier transforms applies for the computation of DTTs as well: from classical radix-2 approaches to mixed-radix algorithms for composite transform sizes, to pruning techniques. All current high-performance libraries can compute the common DTT types (e.g. DCT-II and DCT-IV). FFTW [34] supports the largest variety of DTTs, covering the DCTs and DSTs of types I-IV.

Symmetric convolution

A motivation for considering linear filtering using these transforms is to realize fast convolution in real-valued arithmetic. Another argument is, that the input data might already be in the DTT domain (e.g. DCT-II coefficients of a JPEG image). Unfortunately, none of the DTTs holds the general cyclic convolution property (CCP) (Eq. 2.34). As for the discrete Hartley transform, the transform kernel is not composed from powers of primitive N^{th} roots of unity and hence the transform can not have the CCP [3]. But a point-by-point convolution multiplication property (CMP) [66] of the form

$$x(n) \; \widetilde{\circledast} \; h(n) \quad \circ\!\!-\!\!\bullet \quad \mathcal{T}_c^{-1}\big\{ \, \mathcal{T}_a\{x(n)\} \; \times \; \mathcal{T}_b\{h(n)\} \, \big\} \qquad (2.62)$$

can be derived for special convolution operators, called *symmetric convolution* [66], denoted here by $\circledast$. Eq. 2.62 marks a generalization of the CCP. (Selecting $\circledast = \circledast, \mathcal{T}_{a,b,c} = \mathcal{DFT}$ results at the CCP in Eq. 2.34). Symmetric convolution is particularly suited for filtering impulse responses which inhere symmetries, like for instance linear-phase filters. Filter kernels in image processing often have this property and make DTT-based fast convolution an interesting approach for the 2-D linear filtering of images [67].

Symmetric convolution operators are defined over *symmetric periodic sequences* (SPS), which inhere certain symmetries (whole-sample or half-sample symmetry or anti-symmetry) [66]. Each symmetric convolution operator assumes certain symmetries for the left- and right-hand-side operand, resulting in a large variety of 40 feasible types operators [66]. Each operator is thereby associated with two specific forward transforms $\mathcal{T}_a, \mathcal{T}_b$ and a backward transform $\mathcal{T}_c^{-1}$ (cp. Eq. 2.62). Martucci [66] provides formulations of symmetric convolution for all of the 16 types of discrete trigonometric transforms (eight DSTs and eight DCTs). Several authors proposed methods to realize linear and circular convolution based on symmetric convolution, for instance [66, 131, 49, 81]. These approaches are shortly introduced in the following.

Linear convolution

Martucci [66] proposes a method to realize linear convolution of two sequences $x(n) * h(n)$ using symmetric convolution in the DTT domain—for *all* possible types and combinations of transforms. His approach relies on prior zero-padding of the sequences at both sides (left and right). Considering DFT-based linear convolution, undesirable time-aliasing occurs when the transform is undersized and periodic continuations of the convolved output overlap in time. This defines conditions for the necessary transform size (see Eq. 2.18). A similar phenomenon can occur for DTT-based convolution: Here, the linear convolution is corrupted, when symmetric extensions of the sequences fold back into the range of desired output samples (*fold-back aliasing*). Hence, similar conditions on the transform size exist for DTT-based symmetric convolution [66, 49], with the addition, that proper zero-padding on both sides is required. The different symmetries slightly affect the amount of zero-padding by one to two samples. Ito and Kiya [49] examine linear convolution using the DCT-II as forward and the DCT-I as backward transform $(\mathcal{T}_{a,b} = \mathcal{DCT}\text{-II}, \mathcal{T}_c^{-1} = \mathcal{DCT}\text{-I}$ in Eq. 2.62)

$$s(n) * h(n) =$$

$$\mathcal{R}\Big\{\mathcal{DCT}\text{-I}^{-1}\Big\{\mathcal{DCT}\text{-II}\{\mathcal{P}_1\{x(n)\}\} \times \mathcal{DCT}\text{-II}\{\mathcal{P}_2\{h(n)\}\}\Big\}\Big\} \qquad (2.63)$$

Again, $\mathcal{P}_1, \mathcal{P}_2$ correspond to padding operators and $\mathcal{R}$ to a rectangular window operator, selecting the valid output samples. Ito and Kiya [49] closely review the necessary zero-padding for achieving linear convolution. Their results show that a full $M \times N$-linear convolution using Eq. 2.63 requires at least a transform size of

$$K > \frac{3(M+N-1)}{2} \tag{2.64}$$

The paper includes a very brief comparison of the computational costs to FFT-based linear convolution, stating that their DCT-based method could outperform FFT-based linear convolution. Unfortunately, these parts lack a proper description of the circumstances. Nor are the sizes of operands clearly introduced, neither are actual numbers of operations considered (only runtime models). Their results are picked up later for performance comparisons. Because linear convolution can be realized based on symmetric convolution, the input-partitioning techniques Overlap-Add and Overlap-Save can be applied to DTT-based convolution as well. Zou, Muramatsu and Kiya [131] discuss implementations of partitioned convolution in conjunction with DTT filtering.

Circular convolution

Reju, Koh and Soon [81] derive DTT-formulations of circular convolution by decomposing the DFT into corresponding cosine and sine transforms (using Euler's formula). Their frequency-domain formulation of a N-point circular convolution contains a combination of cosine and sine transforms [81]

$$s(n) \circledast h(n) =$$
$$\frac{1}{4}\breve{C}_1^{-1}\left\{\xi(k)\left(\breve{C}_2\{s(n)\} \times \breve{C}_2\{h(n)\} - \breve{S}_2\{s(n)\} \times \breve{S}_2\{h(n)\}\right)\right\} + \tag{2.65}$$
$$\frac{1}{4}\breve{S}_1^{-1}\left\{\breve{S}_2\{s(n)\} \times \breve{C}_2\{h(n)\} + \breve{C}_2\{s(n)\} \times \breve{S}_2\{h(n)\}\right\}$$

Here, $\breve{C}_1, \breve{S}_1, \breve{C}_2, \breve{S}_2$ correspond to DTTs of symmetrically extended length-$2N$ sequences, decimated by two. Each can be computed using a $2N$-point DTT. Their spectra contain $2N$ real-valued DTT coefficients, from which half are zero. That implies that each spectral convolution in Eq. 2.65, denoted by $\times$, can be realized using N real multiplies. $\xi(k) = \delta(k) + 1$ is a simple scaling factor for the first coefficient. Suresh and Sreenivas [107] further improved the method in [81], by using the DFT of the impulse reponse in combination with DCTs/DSTs (types I and II). Their method saves a DST transform in case of symmetric impulse responses (e.g. linear phase filters). Also the number of operations for the spectral convolution is slightly reduced.

Computational complexity

In the following the DCT-convolution method in [49] is compared to real-data FFT-based convolution (section 2.5.2) in the example of a 256×256-linear convolution. The FFT-convolution is implemented using two 512-point real-to-complex FFTs, one 512-point complex-to-real IFFT [50] and 257 complex-valued multiplications of symmetric DFTs coefficients (cp. section 2.5.2). For simplicity it is assumed that an IFFT consumes the same number of operations as a forward FFT of the same size. In total the FFT approach requires $3 \cdot 7014 + 257 \cdot 6 = 22584$ operations. For the DCT approach (Eq. 2.63) the minimal transform size is $K > 3(256 + 256 - 1)/2$ (Eq. 2.64). Here, the next larger power-of-two $K = 1024$ is selected. All DCT-II transforms are computed using the recent method of Shao and Johnson [94]. Again for simplicity, it is assumed that a DCT-I can be computed with the same number of operations as a DCT-II (in practice it might compute slower [36]). The DCT-convolution [49] requires two 1024-point DCT-II transforms and one 1024-point DCT-I transform to be computed, each costing 18698 operations. The spectrum multiplication makes up for 1024 real-valued multiplications. Altogether this DCT approach results in $3 \cdot 18698 + 1024 = 57118$ operations. This is 2.53 times more than the FFT approach, which demands 22584 operations. Yet with respect to sheer spectral convolutions, the DCT method is faster, consuming only 1024 operations instead of 1542 operations for the FFT approach.

Conclusions

Realizing the spectral convolution by real-valued arithmetic saves some operations over the complex-valued arithmetic, required for the DFT method. This is contrasted by the signficantly increased costs in computing the transforms, whose sizes are also enlarged by the requirement to incorporate further zero-padding to prevent fold-back aliasing. A head-to-head comparison of recent real-data FFT [50] and fast DCT algorithms [94][95] reveals, that an N-point real-data FFT is still faster than an N-point DCT of type II, III, or IV. In practice, this gap widens. The conducted benchmark for FFTW3 [36] disclosed, that DCTs and DSTs compute significantly slower than real-data FFTs (see chapter 7.4). In particular the DCT/DST of type I compute even slower with FFTW [36].

It is concluded, that for real-time audio FIR filtering, DTT-based fast convolution can not outperform DFT-based approaches on general purpose processors. However, DTT techniques might be beneficial, if the input data is already within the DTT domain or when DTT hardware transformers are available, that allow rapid computations of the transforms.

2.5.7. Number Theoretic Transforms

The discrete Hartley transform and discrete trigonometric transforms avoid complex-valued arithmetic, but do not have a general cyclic convolution property (CCP). Moreover, these transforms are still based on cosines and sines, which require a floating-point arithmetic. This section discusses an alternative number theoretic approach which has the CCP and can be realized entirely in integer arithmetic: *number theoretic transforms (NTTs)*. This type of transform was derived in the 1970s as generalized discrete Fourier transform defined over other algebraic structures [75]—in particular finite integer rings or fields. The NTT can be used to implement circular and hence linear convolution in the similar way as with the FFT [4]. Thereby it allows an *exact* convolution, free of round-off errors. NTTs have very elegant implementations in hardware and simplified circuitry. This has been a main area of its applications. Software implementations however are rare. In this respect, the above cited advantages seem not to pay-off equally. Besides their favored properties, NTTs come also along with several implications on the dynamic range, word lengths and filter lengths. Opposed to the FFT, these aspects require detailed attention in applications. The pitfalls are conquered by decomposing larger convolutions into more suitable small convolution (e.g. Agarwal-Cooley algorithm [5] in section 2.6.2). The usage of NTTs is not limited to signal processing. It also serves as a fundamental tool for fast integer multiplication algorithms [87].

This section discusses NTT-based convolution as an alternative to FFT-based convolution in the light of real-time filtering on standard processors. The literature on NTT is exhaustive and quite specific. Only a very brief overview is given. General information on the transform, its properties and implementation can be found in the textbooks by Blahut [13] and Nussbaumer [73]. Details on its historic evolution can be found in [12]. Gudvangen [44] contributed a recent paper discussing its application in audio processing.

Definition

A number theoretic transform (NTT) has the following structure [3, 73]

$$x(n) \quad \overset{\mathcal{N}TT}{\circ\!\!-\!\!\bullet} \quad X(k) = \sum_{n=0}^{N-1} x(n)\,\alpha^{nk} \qquad \mod M \qquad (2.66)$$

$$X(k) \quad \overset{\mathcal{N}TT^{-1}}{\bullet\!\!-\!\!\circ} \quad x(n) = N^{-1} \sum_{k=0}^{N-1} X(k)\,\alpha^{-nk} \qquad \mod M \qquad (2.67)$$

Eq. 2.66 defines a N-point forward NTT and Eq. 2.67 the corresponding inverse NTT. Substituting $\alpha = e^{-2\pi i/N}$ in Eq. 2.66 results in the discrete

Fourier transform (Eq. 2.31). The NTT however is defined over finite rings or fields of integers. All calculations are carried out in $\mathbb{Z}_M = \mathbb{Z}/M\mathbb{Z} = \{0, \ldots, M-1\}$, the set of integers modulo $M \in \mathbb{Z}$.

Negative signal amplitudes as in conventional signed integer quantization are encoded by introducing a positive offset $M/2$. Given that $|x(n)| < M/2$ this representation is unambiguous. The NTT spectrum $X(k)$ also consists of integer numbers $X(k) \in \mathbb{Z}_M$. The NTT coefficients do not have a similar meaning of frequency, magnitude or phase, like for DFT coefficients.

There is not *one* specific NTT, as for instance *the* length-N DFT. The term NTT summarizes many discrete transforms of the structure in Eq. 2.66, 2.67. In order to use a number theoretic transform for fast convolution, it must hold the cyclic convolution property $x(n) \circledast h(n) = \mathcal{NTT}^{-1}\{\mathcal{NTT}\{x(n)\} \times \mathcal{NTT}\{h(n)\}\}$ and the transform must be computable in $\mathcal{O}(N \log N)$. For these requirements to be met, the parameters α, N, M must be chosen according to certain criteria, from which most important conditions are briefly reproduced here. For a thorough review the reader is referred to [3, 73].

- $\alpha^N \equiv 1 \mod M$ must be an order-N root of unity in $\mathbb{Z}_M$. This condition determines possible (maximal) transform sizes N. Given that M is prime, the transform size must hold $N \mid (M - 1)$.

- If the modulus M is composite (non-prime), α must fulfill $\forall n = 1, \ldots, N-1 : \gcd(\alpha^n - 1, M) = 1$ relatively prime.

- The inverse NTT transform in Eq. 2.67 is defined, if the multiplicative inverse N^{-1} exists in $\mathbb{Z}_M$. For prime M this is always the case, as $\mathbb{Z}_M$ is a field.

Due to their common structure, fast NTT algorithms are derived in similar ways like FFT algorithms. For the sake of computational efficiency, the transform length N should be a highly composite number, as for the FFT. Targeting hardware implementations, α should be defined in a way, that its powers α^{nk} can be computed with low complexity (e.g. few digits in their binary representation allowing for simplified circuits). The same holds for the reduction to the modulus M, which in particular should not be odd [3]. Two important classes of NTTs are Mersenne number transforms (MNTs) and Fermat number transforms (FNTs).

Mersenne Number Transforms

By using Mersenne numbers $M = 2^p - 1$ ($p \in \mathbb{N}$ prime) as moduli for the NTT defines the Mersenne number transform (MNT) [3, 73]. The elegance of this transform arises when using simple roots $\alpha \in \{-2, +2\}$. A particular advantage of MNTs are the simple implementation of their arithmetic in hardware.

Additions modulo $2^p - 1$ can be realized by applying little modifications to standard hardware adder units. Multiplications with α^{nk} being powers of two can be realized using bit shifting. Hence, the computation of the transform does not require hardware multiplications units. Unfortunately, the MNT retains the CCP for simple $\alpha \in \{-2, +2\}$ for just two choices of transform lengths $N \in \{p, 2p\}$ [73]. By definition p is prime and hence also $2p$ is little composite (it can only be divided by two). This prevents the application of fast divide-and-conquer algorithms, making the computation of the MNT comparably slow.

Fermat Number Transforms

Fermat primes $M = F_n$ of the form $F_n = 2^{2^n} + 1$ ($n \in \mathbb{N}$) have proven as a better choice for the modulus. Using a Fermat number as the modulus $M = F_n$ for the NTT defines the Fermat number transform (FNT) [3, 73]. Only the first five Fermat numbers are known to be prime: $F_0 = 3, F_1 = 5, F_2 = 17, F_3 = 257, F_4 = 65537$. Comprehensive review of the choice of root α, transform length N and modulus M and their relations can be found in [4], [73] and [44]. In the face of the large extend of the theory, only the most important results are briefly outlined here. The FNT is superior to the MNT in several aspects: It can be shown that the FNT can implement transform sizes $N \mid 2^{n+2}$, which in hardware only require multiplications by powers of two (bit-shifting) and additions. Hence, it is much more flexible in respect to possible transform lengths N. Particularly the fact that it allows highly composite transform sizes (powers-of-two) enabled the use of highly efficient algorithms (e.g. Cooley-Tukey methods). When implementing the FNT on a machine, one has to select a word length for the integer calculations (e.g. $b \in \{8, 16, 32, 64\}$ on general purpose processors). Agarwal and Burrus [4] show, that for the most efficient cases ($\alpha = 2$), the maximal transform is $N_{\max} = 2b$. For example, 32-bit integer arithmetic allows realizing fast circular convolution using the FNT for a maximal transform size of $N_{\max} = 64$ points. For most signal processing applications in acoustic virtual reality this is too short. Several strategies have been proposed to overcome this limitation [11], including Chinese remainder decomposition techniques (see [44]), multi-dimensional index mapping (see 2.6.1 and Agarwal-Cooley algorithm in section 2.6.2) as well as partitioned convolution [4]. Another reason for the need of these decompositions arises from the limitation in dynamic range. For the sake of a correct convolution result, numeric clipping (aliasing in amplitudes) must be avoided, when performing the calculations modulo M. Given a certain word length b this constraint can additionally limit the maximal lengths for the sequences to be convolved, making an additional partitioning necessary [4, 44].

Conclusions

Number theoretic transforms have been an intensive field of research, including a large number of publications over several decades. Yet they did not reach the level of popularity as the FFT. Gudvangen [44] recently discussed the NTTs in relation to audio processing. Whereas several authors considered ASIC and FPGA implementations of the FNT (list of references in [44]), only little is known about its performance in software implementations on current general-purpose processors (desktop processors or mobile processors). Gudvangen [44] concludes that software implementations of the NTT might not be competitive, due to the required modulo arithmetic and unusual word lengths that do not match those typical processors (e.g. 16-/32-/64-bit). Whilst in the 1970s it has been shown that the NTT can outperform an FFT (e.g. [4, 5]), there seems to be no recent study comparing the performance of the NTT in software to established FFT libraries. Also few NTT software implementations are available (for instance the Finite Transform Library (FTL) [22]). It is doubtful that these have the same level of technical maturity as FFT libraries, making a meaningful comparison difficult.

2.6. Number theoretic convolution techniques

Number theory played an important role in the evolution of digital signal processing [68]. Many FFT algorithms and also convolution techniques make use of this mathematical field and its tools. The NTT discussed in the preceding section is an example. A particular important tool is the chinese remainder theorem (CRT), which dates back to the fifth century. Its applications brought up new perspectives on problems, from which many successful algorithms have been derived. These advancements can mostly be found in the form of decomposition schemes. When a large problem instance is decomposed into subsets of easier solvable problems, the CRT is often used to reassemble the complete solution from the partial solutions. An application of the CRT over integer rings are index-mappings. These serve as the basis for many FFT algorithms (e.g. Cooley-Tukey [24], prime-factor algorithm [42, 109]). Burrus [20] showed up their application in directly decomposing DFTs as well as circular convolutions. Circular convolutions of large composite sizes can thereby be split up into small (prime-size) circular convolutions (e.g. a 7-point circular convolution). Nesting these techniques in conjunction with optimal short convolution templates allows gaining efficient algorithms for small to medium sizes (e.g. Agarway-Cooley algorithm [5]). An application of the CRT over polynomial rings is the Winograd short circular convolution algorithm [73]. This algorithm makes use of the polynomial product formulation (see section 2.3) and reduces the polynomial product to lower-degree products in a set of factorial rings. Eventually the complete polynomial is recombining using the CRT.

The aim of this section is to review these decomposition concepts as alternatives to the standard concept of *partitioned convolution*, which is discussed afterwards. The central focus lies on the real-time applicability. The concepts are briefly outlined here, primarily targeting the discussion of the methods considering this aspect. For further details on number theoretic techniques the reader is referred to the textbooks by Blahut [13] and Nussbaumer [73].

2.6.1. Multidimensional index mapping

A technique of fundamental importance in FFT algorithms is *multidimensional index mapping*. The principle idea is to substitute a univariate index variable n by an equivalent mathematical formulation consisting of several variables $n_1, n_2, \ldots, n_P$. Thereby one dimensional indices are mapped to multi-dimensional tuples and vice versa

$$n \leftrightarrow (n_1, n_2, \ldots, , n_P) \tag{2.68}$$

This can be thought of as arranging the elements of a vector in the form of a matrix. In the following the considerations are limited to the two dimensional case $n \leftrightarrow (n_1, n_2)$. A thorough introduction to the technique, including the number theoretic background, can be found in [73, 21, 13]. An essential condition is that the mapping is unique. This can only be achieved if the number of elements $N \in \mathbb{N}$ is not prime and can be factored into $N = N_1 \cdot N_2$ (trivially $N_1, N_2 > 1$). A useful mathematical formalism for such mappings is the linear equation [20, 21]

$$n = \langle K_1 n_1 + K_2 n_2 \rangle_N \quad \text{with constants} \quad K_1, K_2 \in \mathbb{N}_0 \tag{2.69}$$

Two important cases are distinguished. Given that $\gcd(N_1, N_2) = 1$ coprime, a so called *prime-factor map* (*PFM*) can be constructed. This mapping is the basis for the Good-Thomas prime-factor FFT. If N_1 and N_2 have a common factor $(\gcd(N_1, N_2) > 1)$ only a *common-factor map* (*CFM*) is realizable. Cooley-Tukey FFTs are derived by applying this mapping to the DFT in time and frequency [21, 13]. Conditions for uniqueness and the choices of parameters K_1, K_2 are discussed in [20]. Moreover, Burrus [20] pointed out the use of index-mapping for decomposition of circular convolutions and presented how a large one-dimensional circular convolution can be decomposed into several smaller circular convolutions along other dimensions. This makes this decomposition technique worth the consideration as an alternative to partitioned convolution. The results in [20] are outlined and the application to real-time filtering is discussed.

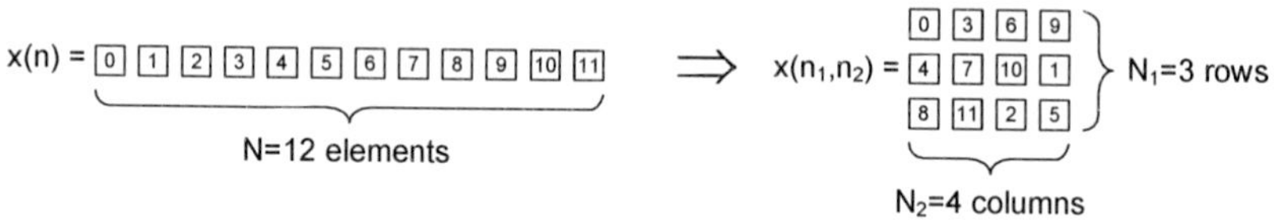

$$x(n) = \boxed{0}\,\boxed{1}\,\boxed{2}\,\boxed{3}\,\boxed{4}\,\boxed{5}\,\boxed{6}\,\boxed{7}\,\boxed{8}\,\boxed{9}\,\boxed{10}\,\boxed{11}$$

N=12 elements

$$\Longrightarrow \qquad x(n_1, n_2) = \begin{array}{cccc}\boxed{0}&\boxed{3}&\boxed{6}&\boxed{9}\\\boxed{4}&\boxed{7}&\boxed{10}&\boxed{1}\\\boxed{8}&\boxed{11}&\boxed{2}&\boxed{5}\end{array}\Big\} \; N_1\text{=3 rows}$$

N_2=4 columns

Figure 2.12.: Example 2-D index map for $N = 12$ elements: $n \equiv 4n_1 + 3n_2$ mod 12 ($N_1 = 3, N_2 = 4, K_1 = 4, K_2 = 3$)

Decomposition of circular convolutions using index mapping

Applying the index mapping Eq. 2.69 to both indices $n \to (n_1, n_2)$ and $k \to (k_1, k_2)$ in the definition of circular convolution Eq. 2.12 results in

$$\tilde{y}(n) = \sum_{k=0}^{N-1} x\langle n - k\rangle_N \cdot h(k) \quad \overset{2.69}{\Longrightarrow}$$

$$\tilde{y}(n_1, n_2) = \sum_{k_1=0}^{N_1-1} \sum_{k_2=0}^{N_2-1} x\langle K_1 n_1 + K_2 n_2 - K_1 k_1 - K_2 k_2\rangle_N \cdot h(K_1 k_1 + K_2 k_2)$$

$$= \sum_{k_1=0}^{N_1-1} \sum_{k_2=0}^{N_2-1} x\langle n_1 - k_1, n_2 - k_2\rangle_N \cdot h(k_1, k_2) \tag{2.70}$$

Eq. 2.70 is a two dimensional $N_1 \times N_2$ convolution. It becomes a true two-dimensional *circular* convolution only if the mapping is cyclic along *both dimensions*. This case is desired, as then a two 2-D circular convolution can be computed using a set of shorter lengths 1-D circular convolutions along the rows respectively columns. Necessary conditions therefore are [20]

$$\gcd(N_1, N_2) = 1 \tag{2.71}$$
$$K_1 = \alpha N_2, \quad K_2 = \beta N_1 \qquad (\alpha, \beta \in \mathbb{N}_0) \tag{2.72}$$

Uniqueness of the mapping furthermore requires that

$$\gcd(\alpha, N_1) = 1 \quad \text{and} \quad \gcd(\beta, N_2) = 1 \tag{2.73}$$

The resulting mapping is a prime-factor map (PFM) [19]. It is an application of the chinese remainder theorem over integer rings [20]. A typical choice is $K_1 = N_2, K_2 = N_1$. This mapping is illustrated in an example for twelve elements in figure 2.12.

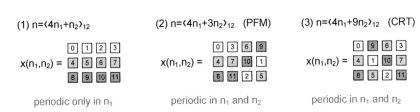

Figure 2.13.: Example 2-D index maps for $N = 12$ elements and their distribution of elements: (1) row-major order, (2) PFM, (3) PFM matching the chinese remainder theorem (CRT)

Applicability for real-time processing

Decomposing large circular convolutions using the reviewed approach could be beneficial for real-time filtering. For instance for linear filtering of short input blocks with long filters, as occuring for real-time FIR filtering with room impulse responses. Also possible, it is computationally inefficient to process such convolutions with large FFT transforms in an unpartitioned way (details are found in chapters 3 and 5). The index-mapping technique can be applied to this problem by interpreting the desired linear convolution as a circular convolution (incorporating zero-padding) and then splitting it into several shorter circular convolutions. The quintessential point is, that for a block-wise real-time filtering, the response on the first input block must be computed *immediately*, in order to avoid unwanted latency.

The index-mapping technique shall be compatible with block-partitioned input sequence. Figure 2.13 illustrates this for the prior example. Ideally, the input blocks (white, light gray, dark gray) become rows or columns in the 2-D representation, as in case (1). Then the 2-D circular convolution can be computed for each row respectively column separately. An initial response of the filter can be computed independently from the subsequent convolutions. Unfortunately, this is not possible. Example (1) is only periodic in one dimension and hence the decomposition in 2.70 cannot be applied. The results would be incorrect.

Figure 2.13 shows two further maps, which are compatible and fulfill the postulated conditions in Eq. 2.71-2.73. Here it can be seen, that original block decomposition gets lost after the mapping. The samples are quasi-chaotically scattered. This has the consequence, that *all* row respectively column circular convolutions have to be performed at once, when the partial filter response of the first input block (white) is required.

Any row or column major layout would require that $K_1 = 1$ or $K_2 = 1$. As $N_1, N_2 > 1$ and $\alpha, \beta > 0$ this is impossible. The index mapping technique in conjunction with circular convolution allows to decompose large convolutions into short convolutions. Therefore, it can be used as a basis for divide-and-conquer algorithms. Unfortunately, it does not allow to distribute the computations over time for real-time processing.

2.6.2. Agarwal-Cooley convolution algorithm

The previously discussed decomposition allows deriving fast convolution techniques directly. Given that the length N is highly composite (many coprime factors), a large convolution can be eventually computed from many tiny circular convolutions. Due to the conditions of the PFM, all of these need to have different and coprime sizes. For example can a circular convolution of length $N = 315 = 3^2 \cdot 5 \cdot 7$ be implemented just using circular convolution of the length 5, 7 and 9. Agarwal and Burrus [3] proposed to compute such short circular convolutions using optimal short convolution templates, which can be derived by hand or using the Toom-Cook algorithm (in section 2.4.1) or the Winograd technique (in the succeeding section 2.6.3). The authors compared the arithmetic complexity to FFT-based convolution. Using optimized short convolutions, their method could save arithmetic operations over the FFT method for small sizes $N < 128$. They also discuss implementations incorporating the NTT. Agarwal and Cooley [5] refined this technique and provided several improvements and combined short and long convolution algorithms. They examined the implementation of longer convolutions using Fermat number transforms (see section 2.5.7). The Agarwal-Cooley convolution algorithm marks an interesting theoretical result, as it requires less operations than a comparable FFT-base convolution for sizes $N < 420$. How these theoretical results translate in practice today is hard to judge. Even the authors [5] pointed out the difficulty of a fair comparison, as the NTT favors specific hardware (cp. conclusions of section 2.6.2). It is doubtful that the algorithm could outperform FFT convolution for the cited sizes on modern general purpose processors. Moreover, the results indicate that FFT convolution seems to be the most performant known technique for longer filters.

2.6.3. Winograd convolution algorithm

Winograd [126] proposed the use of the CRT for polynomial rings for computing circular convolutions. This technique is referred to as the *Winograd (short) circular convolution algorithm* [21, 13]. It provides the tight bound $\Omega(N)$ for the number of multiplications required to compute an N-point circular convolution. Although its number of multiplications is minimal, the number of additions is increased, making the algorithm only efficient for short convolutions [13]. The Winograd algorithm therefore is not of primary interest as an alternative method for computing long convolutions. Its concept is briefly outlined here, aiming the discussion of applicability as a divide-and-conquer scheme for real-time filtering. Details on its derivation and implementation can be found in the textbook by Blahut [13].

Decomposition of circular convolutions using polynomial moduli

An N-point circular convolution can be realized by reducing the corresponding polynomial product modulo $z^N - 1$ (Eq. 2.16 section 2.3)

$$y(n) = x(n) \circledast h(n) \quad \hat{=} \quad Y = X \cdot H \quad \mod M = z^N - 1 \tag{2.74}$$

Now the polynomial modulus $M \in R[z]$ is factored into $Q > 1$ factors $M(z) = M_0(z) \cdots M_{Q-1}$. All factor polynomials M_i have $\deg(M_i) < \deg(M)$, more precisely $\deg(M) = \sum M_i$. Instead of computing product $Y = X \cdot H \mod M$ in $R[z]/M$ directly, the input polynomials (sequences) are first reduced modulo the factors M_i, then products of lower degrees are computed locally, modulo M_i

$$X \equiv X_0, \quad H \equiv H_0 \quad \Rightarrow \quad Y_0 \equiv X_0 \cdot H_0 \qquad \mod M_0 \tag{2.75}$$

$$\vdots$$

$$X \equiv X_{Q-1}, H \equiv H_{Q-1} \quad \Rightarrow \quad Y_{Q-1} \equiv X_{Q-1} \cdot H_{Q-1} \quad \mod M_{Q-1} \tag{2.76}$$

Each line in Eq. 2.75-2.76 corresponds to an independent circular convolution of the form $x_i(n) \circledast h_i(n)$ of length $\deg(M) + 1$. Finally, the complete polynomial Y can be reconstructed (modulo M) from all previously computed $Y_0, \ldots, Y_{Q-1}$ using the CRT for polynomial rings.

The Winograd short circular convolution algorithm is an abstracted technique. Like the Toom-Cook algorithm it is not constructed for actual sequences with given sample values, but derived symbolically. Effectively, it results in a sequence of terms, which allow to evaluate a circular convolution with less terms than the direct formula (Eq. 2.12). The factors of the modulus $M = z^N - 1$ are uniquely determined cyclotomic polynomials [73], which have simple coefficients that are mostly $+1, -1$. Hence the residues modulo M_i in Eq. 2.75-2.76 can be implemented using simple additions and subtractions. The CRT reconstruction can be handled similarly, without multiplications and only using post-additions. Finally, it is remarked, that the products in Eq. 2.75-2.76 do not represent circular convolutions by themselves. It follows from the unique factorization of $z^N - 1$ and the fact that only the first cyclotomic polynomial has the form $z^0 - 1$. Consequently, the algorithm can not be used for decomposing long circular convolutions into shorter circular convolutions.

2.7. Summary

This section reviewed the FFT as a reference method for fast convolution and discussed potential alternatives. The most important fast convolution techniques known today were regarded. The objective was to assess their computational performance and usability for real-time filtering in acoustic virtual reality. The following conclusions are summarized.

Direct convolution is only advised for very short filters. The break-even point where FFT-convolution becomes more efficient was identified with $N \approx 32$ here. This matches prior observations [21, 130]. In practice, the performance of time-domain filters is strongly impacted by vectorization, the memory access pattern and cache utilization [48]. Thus, the exact break-even point can only be found by benchmarking and comparing matured implementations. For shorter filters, interpolation-based methods might be an alternative. Here, nested techniques, like the Karatsuba convolution, can theoretically outperform FFT-based techniques for larger problem sizes and shift the break-even point a little further ($64 < N < 128$). Within the class of transform-based convolution techniques, the FFT could be identified the best method on general purpose computers. This is to a large degree reasoned by today's high performance libraries, which allow for a very fast computation in practice.

Real-valued transforms have been considered in theory to avoid the need of a complex-valued arithmetic. Here, the DHT and the set of DTT have been considered as alternatives to the FFT for implementing linear filtering. Unfortunately, these transforms do not provide simple point-by-point spectral convolutions, as the DFT does. Eventually, it turns out that the complexity is not reduced by real-valued transforms. The conducted benchmark revealed, that the transforms computed significantly slower then comparable real-data FFT. These performance losses over the FFT are substantial, as the transforms itself consume the major part of the computation. Given general purpose processors, there is a clear disadvantage in using these transforms for convolution. This might be different on devices where hardware transformers are available.

An interesting alternative to the DFT could be the NTT, which can be seen as a generalized DFT. Its advantages are a numerically exact convolution entirely based on integers and simplified circuitry when it is implemented in hardware. These have little relevance for acoustic virtual reality realized using of-the-shelf computers. Firstly, round-off errors are not considered problematic in current Virtual Reality (VR) applications. Secondly, as the hardware is given, the required modulo operations in NTTs have to be implemented in software. This might be a substantial drawback with respect to the FFT. Actually, no recent performance comparisons between the NTT and FFT can be found for common processors. Further divide-and-conquer convolution algorithms can be derived from number theory. Multidimensional index-mapping, allows computing large circular convolutions by a set of shorter circular convolutions. Unfortunately, these decomposition result in unfavorable dependencies, making it impossible to stretch the computations over time. Hence, the technique cannot serve as an alternative to partitioning, which is reviewed in the next chapter.

2. Fast convolution techniques

3. Partitioned convolution techniques

Partitioned convolution methods split the input samples and/or the filter impulse responses into *blocks* and define a semantically equivalent convolution on the basis of sub convolutions of these parts. These techniques are mostly applied to divide linear convolutions. A *partition* is thereby commonly associated with decomposition into *blocks*, i.e. connected sequences of samples or filter coefficients that are neighboring in time. In several publications the term *segmented convolution* is used as a synonym for partitioned convolution. This work distinguishes between these two terms. Here, a segment is considered as a uniform sub partition (see chapter 6).

Partitioned convolution is a divide-and-conquer technique. Particularly in real-time processing, it is not only beneficial, but also necessary to split a large linear convolution into several shorter ones. As input signals are of indefinite lengths here, they need to be processed in blocks of samples. The availability of input blocks thereby depends on the actual runtime. This requires realizing the FIR filtering *block by block* in the form of *running convolutions*. Aiming at a low processing latency, partial output results of filtering procedure have to be provided in the meantime. By making use of the well-known Overlap-Add (OLA) or Overlap-Save (OLS) scheme, all previously discussed fast convolution techniques can be adapted to suit this purpose. However, remarkable computational saving can also be achieved by splitting the filter into several sub filters. In order to fully exploit these circumstances, specific partitioned convolution algorithms are needed. These are examined in chapter 5 and 6. In comparison to a large, unpartitioned filter, the set of shorter sub filters can be realized with a much higher computational efficiency. The computational savings are immense.

Many of the previously discussed fast convolution algorithms found on the divide-and-conquer principle and compute longer convolutions by decomposing them into shorter ones. This raises the question, if these concepts can be directly applied to derive fast real-time filtering techniques. *Block-wise* decompositions are favored in real-time processing, because they allow distributing the computation over the runtime, with respect to the availability of input blocks and requirements of intermediate output blocks. Number theoretic strategies, for instance decomposing circular convolutions using multidimensional index mapping (see section 2.6.1 or the Winograd algorithm), do not hold this property. Some interpolation-based techniques make implicit use of partitioning the sequences. The Karatsuba convolution algorithm is real-time capable [47].

This chapter introduces the basic principles of partitioned convolution, which can be applied in two ways: input partitioning and filter partitioning. The OLA and OLS technique for computing running convolutions are presented. Filter partitions are formally defined and the resulting filter structure is introduced. Finally, a classification of convolution algorithms is presented, which marks the guideline for the subsequent chapters of this thesis.

Motivation

For a transform-based linear convolution it would be desirable, that the input sequences of lengths M and N could be represented with an identical number of elements in the frequency-domain (mathematically speaking a bijection): for example, realizing a 1024×128 linear convolution using a 1024-point and a 128-point DFT spectrum. For each sequence on its own this is possible (e.g. real-data DFT in section 2.5.2). However, due to the mathematical background (diagonalization of circulant square matrices, see section 2.5.1), this is not possible. The discrete spectra must be compatible and of the same size N. That enforces zero-padding to achieve linear convolution as desired for audio filtering (see section 2.3).

FFT-based linear convolution methods deliver the highest efficiency (least computation per filtered output sample), when the input signal and filter have matching lengths $M \sim N$. This case is in this work referred to as *balanced* convolution problem. Here, viable transform sizes $K \geq M+N-1 \approx 2M = 2N$ can be chosen close to the sequence lengths M, N. As a result, the amount of zero-padding is minimal and the sequences are represented by the shortest possible DFT spectra with a minimal number of DFT coefficients.

If both lengths M, N strongly differ ($M \gg N$ or $M \ll N$), comparably large transforms must be computed of relatively short sequences, involving extensive zero-padding. Moreover, the DFT spectra are bloated, increasing the computation for spectral convolutions unnecessarily. Considering the example of filtering a three-minute long audio file with a room impulse response of 1-second duration, these disadvantages become clear. Given a sampling rate of 44.1 kHz, two sequences of lengths $M = 3 \cdot 60 \cdot 44100 = 7938000$ and $N = 44100$ needed to be convolved. This would not only require the computation of a very large FFT, which is rather cache-inefficient. Also would the 1 s-filter have to be padded with at least $7938000-1$ zeros.

Using partitioned convolution, a large *unbalanced* convolution problem can be broken up into several smaller *more balanced* convolutions, which can be computed with much higher computational efficiency on their own. The approach is beneficial for both, offline and real-time convolution. Moreover, the technique is not limited to transform-based convolution. But particularly transform-based convolution can heavily exploit these schemes by reusing many previously transformed spectra.

Partitioning can be applied to the input signal, as well as the filter impulse response. In the following these two cases are reviewed in more detail.

3.1. Input partitioning

The first approach is partitioning the input signal. Let $x(n)$ be a signal of (potentially) indefinite length ($n \geq 0$). A *uniform* partitioning of $x(n)$ is a decomposition into a sequence of finite-length sub signals $x_0(n), x_1(n), \ldots$ called *blocks*, which all have an equal *block length* B

$$x(n) = \sum_{i \geq 0} x_i(n - Bi)$$

$$(0 \leq n < B, i \geq 0) \qquad (3.1)$$

$$x_i(n) = \mathcal{R}_B\{ \, x(n + Bi) \, \}$$

Given that $x(n)$ has a finite length M, there are $\lceil \frac{M}{B} \rceil$ blocks $x_i(n)$. If M is not a multiple of B, the final block is zero-padded to match the length B. Audio streams in real-time signal processing have indefinite lengths and thereby an indefinite number of blocks. Nevertheless, a distinct first block $x_0(n)$ always exists, at the point in time that the streaming started.

Real-time running convolutions are always computed using a *uniform input partitioning*, because in real-time audio streaming the block length is fixed and does not vary over time. Fast *offline* convolution could theoretically benefit from differently sized input blocks. However, many transform-based fast convolution algorithm gain their efficiency from reusing the spectra of already transformed blocks. Thereby, a uniform partitioning of signal and filter allows the largest extent of reusing spectra, which are all of the same size. These relations will be more closely reviewed in chapter 5. Non-uniformly partitioned input signals, with block lengths that vary over time, are uncommon in audio processing.

3.1.1. Running convolutions

The linear convolution of a partitioned input signal as in Eq. 3.1 can be implemented using unpartitioned convolutions with the help of two well-known techniques: the Overlap-Add (OLA) and Overlap-Save (OLS) method [21, 74]. Both methods on their own are first of all independent from a

$$
\begin{bmatrix} y_0 \\ y_1 \\ y_2 \\ y_3 \\ y_4 \\ y_5 \\ y_6 \\ y_7 \\ y_8 \\ y_9 \\ y_{10} \\ y_{11} \end{bmatrix}
=
\begin{bmatrix}
h_0 & 0 & 0 & 0 & 0 & 0 & 0 & 0 \\
h_1 & h_0 & 0 & 0 & 0 & 0 & 0 & 0 \\
h_2 & h_1 & h_0 & 0 & 0 & 0 & 0 & 0 \\
h_3 & h_2 & h_1 & h_0 & 0 & 0 & 0 & 0 \\
h_4 & h_3 & h_2 & h_1 & h_0 & 0 & 0 & 0 \\
0 & h_4 & h_3 & h_2 & h_1 & h_0 & 0 & 0 \\
0 & 0 & h_4 & h_3 & h_2 & h_1 & h_0 & 0 \\
0 & 0 & 0 & h_4 & h_3 & h_2 & h_1 & h_0 \\
0 & 0 & 0 & 0 & h_4 & h_3 & h_2 & h_1 \\
0 & 0 & 0 & 0 & 0 & h_4 & h_3 & h_2 \\
0 & 0 & 0 & 0 & 0 & 0 & h_4 & h_3 \\
0 & 0 & 0 & 0 & 0 & 0 & 0 & h_4
\end{bmatrix}
\cdot
\begin{bmatrix} x_0 \\ x_1 \\ x_2 \\ x_3 \\ x_4 \\ x_5 \\ x_6 \\ x_7 \end{bmatrix}
$$

Figure 3.1.: Overlap-Add convolution illustrated as a matrix product [21]

filter partitioning. They can be used with linear convolution, but are mostly applied in conjunction with circular convolution (e.g. FFT techniques). Here, only the basic principles are briefly summarized. The two techniques are later combined with FFT-based implementations in section 4.1 in chapter 4 and reviewed in detail.

Overlap-Add

The Overlap-Add (OLA) method originates from the input partitioning. Each length-B input block $x_i(n)$ is convolved with the length-N impulse response $h(n)$, producing overlapping partial output signals $y_i(n) = x_i(n)*h(n)$ of the length $B+N-1$. As the latter exceeds the block length $(B+N-1 > B)$, the overlapping $N-1$ samples, must be *added* to the final result $y(n)$. This leads to the name Overlap-Add. The method can be visualized using the matrix-vector product formulation of linear convolution [21]. Figure 3.1 shows an example of an 8×5 linear convolution. Thereby, the Overlap-Add method splits the convolution matrix vertically [21]. These sub matrices can then be multiplied with the sub vectors of the input. The sub matrices contain 4×5 convolution matrices and additional blocks of zeros (above or below), which realize the delay. Here, the two partial length-7 results overlap in three samples (marked dark gray). Note, that the output is not partitioned. Technically, the full result can be computed using two 4×5 linear convolutions plus three additions for the overlap.

$$
\begin{bmatrix} y_0 \\ y_1 \\ y_2 \\ y_3 \\ y_4 \\ y_5 \\ y_6 \\ y_7 \\ y_8 \\ y_9 \\ y_{10} \\ y_{11} \end{bmatrix} = \begin{bmatrix} h_0 & 0 & 0 & 0 & 0 & 0 & 0 & 0 \\ h_1 & h_0 & 0 & 0 & 0 & 0 & 0 & 0 \\ h_2 & h_1 & h_0 & 0 & 0 & 0 & 0 & 0 \\ h_3 & h_2 & h_1 & h_0 & 0 & 0 & 0 & 0 \\ h_4 & h_3 & h_2 & h_1 & h_0 & 0 & 0 & 0 \\ 0 & h_4 & h_3 & h_2 & h_1 & h_0 & 0 & 0 \\ 0 & 0 & h_4 & h_3 & h_2 & h_1 & h_0 & 0 \\ 0 & 0 & 0 & h_4 & h_3 & h_2 & h_1 & h_0 \\ 0 & 0 & 0 & 0 & h_4 & h_3 & h_2 & h_1 \\ 0 & 0 & 0 & 0 & 0 & h_4 & h_3 & h_2 \\ 0 & 0 & 0 & 0 & 0 & 0 & h_4 & h_3 \\ 0 & 0 & 0 & 0 & 0 & 0 & 0 & h_4 \end{bmatrix} \cdot \begin{bmatrix} x_0 \\ x_1 \\ x_2 \\ x_3 \\ x_4 \\ x_5 \\ x_6 \\ x_7 \end{bmatrix}
$$

Figure 3.2.: Overlap-Save convolution as a matrix product [21]

Overlap-Save

The Overlap-Save (OLS) method is designed the other way round: It starts by partitioning the output signal and then formulating appropriate convolutions to obtain these non-overlapping output blocks from overlapping ranges of input samples. Again, the method can be nicely illustrated using the matrix-vector product [21], as shown in figure 3.2. The Overlap-Save method splits the convolution matrix horizontally. Each output block results from the multiplication of a sub matrix with the (unpartitioned) input vector. As the output blocks do not overlap, the extra additions as for the OLA method fall apart. Each output block is composed from the turn-off transient of the preceeding input block and the turn-on transient of the current input block. Considering an N-tap direct-form FIR filter (as in figure 3.3), the turn-on transients are produced from the new input samples, while the turn-off transients result from previous $N - 1$ input samples, which are still stored in the accumulators.

In the OLS method, the sub convolutions are defined on overlapping regions of the input samples. Other than for the OLA method, the sub matrices here do form linear convolution matrices. The output blocks are obtained by computing three 8×5 linear convolutions and by selecting only the last four samples of each sub convolution. The name of the method results from the fact, that only a subset of outputs from each sub convolution are *stored* or *saved* into the final output.

x(n)

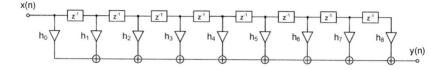

Figure 3.3.: An unpartitioned 9-tap direct-form FIR filter

3.2. Filter partitioning

A second approach is to decompose a large FIR filter by partitioning its *impulse response*. This is referred to as *filter partitioning*. Opposed to input partitioning, which is usually predefined and fixed, filter partitioning can be realized in many different ways and can be thought of as a *parameter*. The way the filters are partitioned, defines different classes of convolution algorithms.

3.2.1. Filter partitioning scheme

The principle of filter partitioning is illustrated in figure 3.6. A length-N filter impulse response is decomposed into P sub filters of length $N_0, \cdots, N_{P-1}$. It is claimed, that all sub filters cover the length of the original filter

$$\sum_{i=0}^{P-1} N_i = N \tag{3.2}$$

Sometimes the strict match ($=$) is relaxed to simple coverage ($\geq$) incorporating zero-padding

$$\sum_{i=0}^{P-1} N_i \geq N \tag{3.3}$$

Each sub filter impulse response $h_i(n)$ relates to a connected interval $n_i^{\text{first}} \leq n \leq n_i^{\text{last}}$ of filter coefficients in the filter h. All sub filters are neighboring in time

$$\forall 0 \leq i < P-1 : \ n_{i+1}^{\text{first}} = n_i^{\text{last}} + 1 \tag{3.4}$$

This implies, that all sub filters must adjoin and that no filter coefficients can be left out by the partition. In other words, the partition is a decomposition into blocks. For convenience two variables are defined.

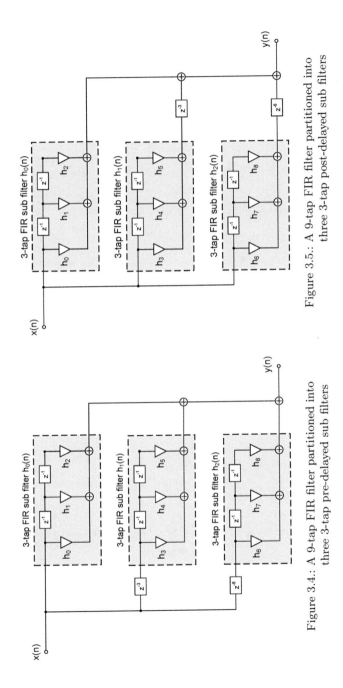

Figure 3.5.: A 9-tap FIR filter partitioned into three 3-tap post-delayed sub filters

Figure 3.4.: A 9-tap FIR filter partitioned into three 3-tap pre-delayed sub filters

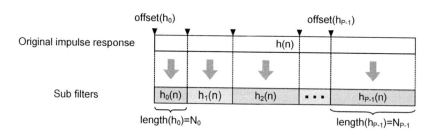

Figure 3.6.: General filter partitioning scheme

The *sub filter length* (the number of filter coefficients) is denoted by

$$\textbf{length}(i) := N_i = n_i^{\text{last}} - n_i^{\text{first}} + 1 \tag{3.5}$$

The position of the sub filter within the original impulse response is

$$\textbf{offset}(i) := n_i^{\text{first}} = \sum_{j<i} N_j \tag{3.6}$$

A sub filter impulse response $h_i(n)$ is obtained by extracting N_i coefficients from the filter $h(n)$

$$h_i(n) = \mathcal{R}_{N_i}\{\ h(n + \text{offset}(i))\ \} \tag{3.7}$$

The original filter $h(n)$ is assembled from all sub filters using the relation

$$h(n) = \sum_{i=0}^{P-1} h_i(n - \text{offset}(i)) \tag{3.8}$$

For reasons which are explained in chapter 6, only ordered decompositions are meaningful, in which the sub filter lengths form a non-decreasing sequence

$$N_0 \le N_1 \le \cdots \le N_{P-1} \tag{3.9}$$

A filter partition, symbolized by $\mathcal{P}$, can be formally written down as a sequence of sub filter lengths

$$\mathcal{P} = (\ N_0, \ldots, N_{P-1}\ ) \tag{3.10}$$

Sub filter lengths $\forall i : N_i \in \mathcal{S}$ are selected from a set of feasible sub filter lengths $\mathcal{S}$. It is usually determined by the convolution algorithm and mostly a true subset of the natural numbers, for instance powers-of-two with certain limits.

Classification of filter partitions

A filter is called *unpartitioned* when $P = 1$. If a partition consists of multiple sub filters of the same size $\forall i, j : N_i = N_j$, it is called a *uniform partition*. Otherwise, it is called a *non-uniform partition*. Due to the order in Eq. 3.9, any non-uniform partition is assembled from uniform sub partitions. These are referred to as *segments* of the partition.

3.2.2. Filter structure

Inserting the block decomposition in Eq. 3.8 into the linear convolution in Eq. 2.3 yields

$$
\begin{aligned}
y(n) \quad &= \quad x(n) * h(n) \\
&\overset{3.8}{=} \quad x(n) * \left[\sum_{i=0}^{P-1} h_i(n - \text{offset}(i)) \right] \quad &(3.11) \\
&= \quad \sum_{i=0}^{P-1} \left[[x(n) * \delta(n - \text{offset } h_i(n))] * h_i(n) \right] \quad &(3.12) \\
&= \quad \sum_{i=0}^{P-1} \left[[x(n) * h_i(n)] * \delta(n - \text{offset } h_i(n)) \right] \quad &(3.13)
\end{aligned}
$$

Writing the delays in Eq. 3.11 as convolutions with time-shifted unit impulses yields Eq. 3.12. As convolution is commutative, the delays can be shifted as in Eq. 3.13.

Eq. 3.13 defines the typical block diagram of partitioned FIR filters. This is illustrated with an example, by partitioning 9-tap FIR filter in figure 3.3 into three 3-tap filters. The resulting partitioned FIR filter networks are depicted in Fig. 3.4-3.5. All sub filters are arranged in a parallel structure and are fed with the same input signal. The total output is added up from all sub filter outputs. Each sub filter branch must be delayed according to the offset of the sub filter in the whole impulse response. The necessary delays for each sub filter branch can be realized in various ways: pre-delays (as in Fig. 3.4), post-delays (as in Fig. 3.5) or combinations of both. Correctness holds as long as each sub filter branch accumulates its designated delay, according to the offset of its associated filter part. In software, delays are realized by read or write positions on buffers.

For the sake of clarity, the decomposition concept is visualized here using time-domain FIR filters. The true potential filter partitioning is exploited by realizing the sub filters with appropriate fast convolution techniques.

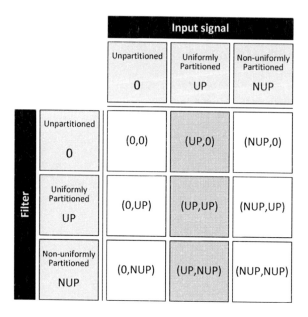

Figure 3.7.: Classes of convolution algorithms with and without partitioning

3.3. Classification of partitioned convolution algorithms

Guiding the further study of algorithms in this thesis, the author introduces a classification of convolution methods. Three different types of partitions were introduced in section 3.2.1. An unpartitioned filter, denoted by **0**. A uniformly partitioned and a non-uniformly partitioned filter, abbreviated by **UP** and **NUP**. Applying these three types for both operands, the signal and the filter, results in a set of nine combinations which are listed in figure 3.7. Each class is written as a pair $(\cdot_S, \cdot_F)$, where the first element corresponds to the partitioning of the signal and the second to the partitioning of the filter. This classification groups algorithms by their use of a block-wise splitting of the operands. Many other classifications are imaginable, for instance based on the fast convolution paradigm or algorithmic complexity classes. The choice depends on the objective and point and view. Here, real-time processing is in the focus, for which the partitioning marks a cornerstone and leads to quite different algorithmic approaches.

The class $(0_S, 0_F)$ summarizes methods which do not facilitate a block-decomposition, neither of the signal nor of the filter. This includes CRT-based techniques, like the Winograd convolution algorithm (section 2.6.3)

and the PFM index-mapping technique (section 2.6.1). Considering the transform algorithms as black boxes, the class also includes all basic (unpartitioned) transform-based methods in chapter 2.

Block-based real-time audio processing introduces a uniform partitioning of the input signals. All relevant methods for this area are hence found in the second column of figure 3.7 (marked gray). Conversely, that does not imply that all methods within these classes are actually real-time capable. Not all Toom-Cook algorithms (section 2.4.1) suit low-latency real-time processing [47]. An exception is the Karatsuba algorithm which belongs to the class $(\mathrm{UP_S}, \mathrm{UP_F})$.

The research in this thesis focuses on the three classes $(\mathrm{UP_S}, \mathrm{0_F})$, $(\mathrm{UP_S}, \mathrm{UP_F})$ and $(\mathrm{UP_S}, \mathrm{NUP_F})$ in conjunction with transform-based fast convolution. FFT-based convolution algorithms for each class are presented. Their properties for real-time filtering in acoustic virtual reality are discussed. The analysis of their computational complexity is of particular interest. The runtime complexity classes are derived.

Everything besides the central column in figure 3.7 is not suitable for real-time filtering with a low latency. However, methods in these classes can be useful for offline filtering, where latency is irrelevant. Due to the commutativity of convolution, signal and filter can be arbitrarily swapped, if it is beneficial. In this respect, only three classes have not been considered so far: $(\mathrm{0_S}, \mathrm{0_F})$, $(\mathrm{0_S}, \mathrm{NUP_F})$, $(\mathrm{UP_S}, \mathrm{UP_F})$ and $(\mathrm{NUP_S}, \mathrm{NUP_F})$. From the study of the real-time methods in the next chapters it follows, that non-uniform partitions are only beneficial in the context of real-time processing, where a processing with low latency is desired. Even if the latency is irrelevant (offline convolution), a partitioning in both operands is mostly still advantageous (e.g. memory locality, improved cache utilization). These offline filtering techniques perform most efficient, when the reuse of previously computed spectra is maximized. Hence, the class $(\mathrm{UP_S}, \mathrm{UP_F})$ covers also these applications. However, as offline convolution lies not in the focus of this work, a detailed study of optimal parameters for these cases is not conducted.

4. Elementary real-time FIR filtering using FFT-based convolution

This section discusses the application of FFT-based fast convolution for real-time FIR filtering. In order to process the consecutive blocks of the input stream, the basic FFT convolution method (introduced in section 2.5.2) is combined with the Overlap-Add and Overlap-Save schemes (see section 3.1.1). This chapter reviews methods which belong to the class $(\mathrm{UP_S}, 0_F)$ in figure 3.7. They do not partition the filter impulse response and process it as a whole. Firstly, different algorithms for time-invariant filtering are introduced. Then their computational complexity is analyzed and the limitations of methods with unpartitioned filters are identified. Important aspects like time-variant filtering or multiple inputs or outputs are first developed at these conceptually simple techniques here. Later they are extended to uniformly and non-uniformly partitioned convolution methods.

4.1. FFT-based running convolutions

In the following a real-time FIR filtering problem is regarded, as it has been defined in section 1.3. A single channel audio stream $x(n)$, provided and processed in length-B blocks $x_0(n)$, $x_1(n)$, $\ldots$, is filtered with an N-tap impulse response $h(n)$. The transform size K is introduced as a third parameter. Two separate workflows are distinguished. The *filter processing* involves all steps to convert the time-domain impulse response into a frequency-domain representation. *Stream processing* describes all computations filtering one input block, producing one output block. The algorithmic complexity is regarded in two measures

- $T^{\mathrm{stream}}(B, N, K)$ marks the computational costs for filtering a *single sample* of the stream.

- $T^{\mathrm{ftrans}}(B, N, K)$ states the costs for transforming the *complete* length-N filter, so that it can be used in the corresponding convolution algorithm.

Depending on the type of examination, both cost functions can be purely analytic (theoretic number of arithmetic operations) or based on measured performance data (average number of CPU cycles on the target machine, see chapter 7).

4.1.1. Overlap-Add

Firstly, the FFT-based running convolution using the Overlap-Add scheme is presented. The resulting algorithm is visualized in figure 4.1. It works as follows

- The next length-B input block is zero-padded to length-K and transformed using a K-point FFT .

- The length-N impulse response is zero-padded to length-K and transformed using a K-point FFT.

- The input and filter DFT coefficients are pair-wisely multiplied (spectral convolution)

- The result is transformed back into the time-domain using a K-point IFFT. It forms a partial convolution result of length $B + N - 1$, which is buffered.

- The length-B output block is added up from the overlapping partial results (Overlap-Add step).

The transform size K must be chosen sufficiently large to avoid time-aliasing (cp. Eq. 2.18)

$$K \geq B + N - 1 \tag{4.1}$$

For the sake of clarity, all partial output signals are stored separately in figure 4.1. In actual implementations, all partial outputs are accumulated in a single output buffer. It must be large enough to store at least $B + N - 1$ samples. Leaving the filter transformation out of consideration, the algorithm has the following computational cost per filtered output sample

$$T_{\text{OLA-FFT}}^{\text{stream}}(B, N, K) := \frac{1}{B}\Big[T_{\text{FFT}}(K) + T_{\text{CMUL}}(K) + T_{\text{IFFT}}(K) + \tag{4.2}$$
$$T_{\text{ADD}}(B + N - 1) \Big]$$

Note, that the number of extra additions depends on the filter length N. For long filters they can cause a significant computational overhead.

4.1.2. Overlap-Save

Figure 4.2 illustrates the corresponding Overlap-Save algorithm. The method is similar to the OLA approach in the way the filter is transformed and the spectral convolution is computed. But it differs in the following aspects:

- The input FFT is computed from a K-point sliding window of the input. Before the transformation, the previous contents are shifted B samples to the left and the next length-B input block is stored rightmost.

- From the output of the K-point IFFT, the $K - B$ leftmost samples are time-aliased and therefore discarded. The B rightmost samples are saved into the output block.

The periodicity of the transform allows to implement the OLS method in different ways. The algorithm can be modified by circular shifting of the buffers (input or filter), altering the positions where data is written into the input buffer and filter buffer and read from the output buffer. Shift operations on the sliding window can be minimized by using larger ring buffers for the input samples. Then actual shifts (copy operations) only have to be performed in the event of buffer wraps. By cyclic shifting of the filter buffer, the valid output block can be moved to the leftmost position. This can be an advantage, when partial inverse transforms shall be computed.

The computational complexity per output sample of the OLS algorithm is

$$T_{\text{OLS-FFT}}^{\text{stream}}(B, N, K) := \frac{1}{B}\Big[T_{\text{FFT}}(K) + T_{\text{CMUL}}(K) + T_{\text{IFFT}}(K) \Big] \quad (4.3)$$

The filter transformation consists of a single K-point FFT for both methods

$$T_{\text{OLA-FFT}}^{\text{ftrans}}(B, N, K) = T_{\text{OLS-FFT}}^{\text{ftrans}}(B, N, K) := T_{\text{FFT}}(K) \quad (4.4)$$

In principle, both methods require the same number of FFTs, IFFTs and complex-valued multiplications. A slight advantage of the OLS over the OLA approach is, that it avoids the extra additions for the overlapping samples.

4.1.3. Computational complexity

The complexity classes of the methods are analyzed by inserting the theoretical costs of the operations (see section 7.4.1) into the cost functions defined above. Therefore, the transform size is chosen the smallest possible value $K = B+N+1$ (Eq. 4.1). FFTs and IFFTs of length K are assumed to have a runtime of the general form $kK \log_2 K$, with a constant $k \in \mathbb{R}^+$ (details see section 7.4).

$$T_{\text{OLS-FFT}}^{\text{stream}}(B, N, K = B+N+1) = \frac{1}{B}\Big[2kK \log_2 K + 6K \Big]$$
$$= \frac{1}{B}\Big[2k(B+N+1) \log_2(B+N+1) + 6(B+N+1) \Big] \quad (4.5)$$

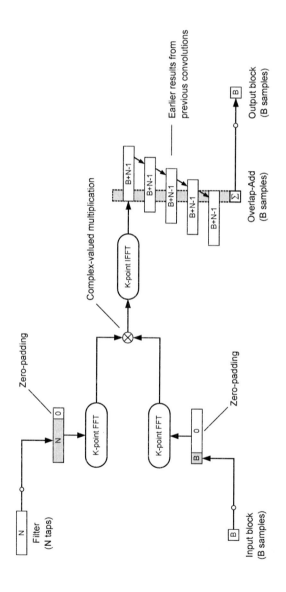

Figure 4.1.: Overlap-Add (OLA) running convolution using the FFT

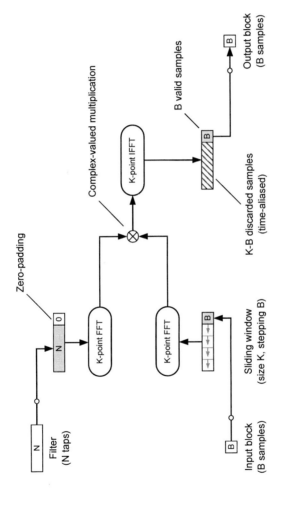

Figure 4.2.: Overlap-Save (OLS) running convolution using the FFT

The asymptotic runtime is now regarded as a function of the filter length N. Assuming the block length B to be a fixed constant, Eq. 4.5 reduces to the functional form $\epsilon_1((N + \epsilon_2) \log_2(N + \epsilon_2) + \epsilon_3 N)$ with constants $\epsilon_1, \epsilon_2, \epsilon_3$. Without formal proofs, the following complexity classes hold for a constant block length B

$$\left. \begin{array}{l} T_{\text{OLA-FFT}}^{\text{stream}}(B \text{ const.}, N, K = B + N + 1) \\[4pt] T_{\text{OLS-FFT}}^{\text{stream}}(B \text{ const.}, N, K = B + N + 1) \\[8pt] T_{\text{OLA-FFT}}^{\text{ftrans}}(B \text{ const.}, N, K = B + N + 1) \\[4pt] T_{\text{OLS-FFT}}^{\text{ftrans}}(B \text{ const.}, N, K = B + N + 1) \end{array} \right\} \in \mathcal{O}(N \log N) \qquad (4.6)$$

This result is counter intuitive. FFT-based convolution is widely favored over time-domain filtering because it is significantly faster. However, the latter has a runtime complexity per filtered sample of $\mathcal{O}(N)$ (Eq. 2.10). From $\mathcal{O}(N) \subset \mathcal{O}(N \log N)$ it follows, that a filter length N must exist, for which time-domain filtering becomes more efficient than the presented basic FFT-based approach. This break-even point is enormously large and way beyond all practical problem sizes. For a realistic FFT constant of $k = 1.7$, the number of operations in Eq. 4.5 exceeds Eq. 2.10 for filter lengths $N > 10^{23}$. Note, that this result only holds for FFT-based convolution *without* filter partitioning. Methods, that partition the filter, break these boundaries and their complexity per filtered sample lies in the class $\mathcal{O}(N)$ or even lower (see chapters 5 and 6).

The filtering costs of the Overlap-Add approach lie in the same class $\mathcal{O}(N \log N)$, as the extra additions are only linear in N. Trivially, the filter transformation for both algorithms has the runtime complexity of the FFT. These complexity classes hold as well for implementations with all other fast transforms introduced in chapter 2, which are computable in $\mathcal{O}(N \log N)$ time.

Eq. 4.5 indicates that the computational costs per filtered sample are nearly anti-proportional to the block length B for large N. In other words, doubling the block length will approximately halve the costs. Note, that the latency is not determined by the block length alone, but also by buffering and the hardware. Accepting more latency generally lowers the computational costs.

4.1.4. Transform sizes

The choice of transform size for the FFTs is critical for the performance of both algorithms. Simply taking the smallest possible size $K = B + N - 1$ will diminish the performance in the majority of cases (see section 7.6).

A common choice is selecting the next larger power-of-two

$$K = \min\{ \, 2^i \geq B + N - 1 \mid i \in \mathbb{N} \, \} \qquad (4.7)$$

The best results are achieved by optimizing the transform K to minimize the computational costs

$$K_{\mathrm{opt}} = \underset{K}{\mathrm{argmin}}\{ \, T(B, N, K) \mid K \geq B + N - 1 \, \} \qquad (4.8)$$

where $T(B, N, K)$ is a cost function (model) of the considered algorithm. Given a performance profile of a given machine (see chapter 7), Eq. 4.8 is evaluated for all considered transform sizes (e.g. 1–65.536) and the optimal transform size K_{opt} is found. On current computers, this optimization procedure is usually a matter of milliseconds. Depending on the application, the costs of the stream processing or filter processing can be selected. Mostly the costs of the stream processing (Eq. 4.2 and 4.3) are regarded, as a filtering free of dropouts has higher importance than faster filter updates. The influence of the transform size is deeper examined in the following section.

4.1.5. Performance

The computational costs of both algorithms are examined for different combinations of block lengths B (corresponding to latencies) and filter lengths N. The cost functions in Eq. 4.2, 4.3 and 4.3 are evaluated with the performance profile of the test system (see chapter 7). Real-data transforms are used and only complex-conjugate symmetric DFT coefficients are stored and processed, yielding the modified cost function

$$T_{\mathrm{OLS\text{-}RFFT}}^{\mathrm{stream}}(B, N, K) := \frac{1}{B} \Big[\, T_{\mathrm{FFT\text{-}R2C}}(K) + T_{\mathrm{CMUL}}\Big(\Big\lceil \frac{K+1}{2} \Big\rceil \Big) + $$
$$T_{\mathrm{IFFT\text{-}C2R}}(K) \, \Big] \qquad (4.9)$$

Figure 4.3 shows the costs for the streaming filtering over the filter length N, for a fixed block length of $B = 128$ samples. The costs are measured as the (average) number of CPU cycles per filtered output sample. Solid lines correspond to optimized transform sizes (Eq. 4.8) and dotted lines to power-of-two transform sizes (Eq. 4.7). The latter become comparably inefficient, when $B + N - 1$ just exceeds a power-of-two. By considering arbitrary FFT sizes a relatively even curve of the computational costs is achieved. However, the curves are not exactly smooth. This is not measurement noise, but the results of the the the fact, that, depending on the actual transform size, very different decomposition strategies are employed within the FFTs. For small filter lengths, the difference between OLA and OLS is marginal. Towards longer filters it becomes significant. Figure 4.4 shows the distribution of computational costs for the OLS algorithm. It can be clearly seen, that the major computation is spent on fast Fourier transforms, whereas the spectral convolution consume only a fraction of the runtime. This allows the conclusion, that major savings are achieved on side of the transforms. The spectral convolutions have a minor impact on the algorithms' performance.

Figure 4.5 depicts costs per output sample for the OLS method for the block lengths 128, 256, 512 and 1024. Again, the costs are derived from the benchmarked data of the test system (see chapter 7). As expected, larger block lengths (corresponding to more latency) reduce the computational effort for the filtering. In order to identify the magnitude in savings, the relative speedups (reduction in costs), when changing from a short reference block length $B = 128$ a large block length, are plotted in figure 4.6. Doubling the block length from $B = 128$ to $B = 256$ samples nearly halves the costs. A four times larger block length $B = 512$ demands a little bit less than a quarter of the effort. An eight-fold block length $B = 1024$ reduces the computation by about a factor of 7-8. These ratios hold for sufficiently long filters $N > 6000$. For shorter filter lengths the savings for larger block lengths are less articulated. The observations show a good agreement with the theoretical results in section 4.1.3.

While the block length B showed a strong impact on the computational load per filtered sample, it does only marginally affect the cost for transforming the filters into the frequency-domain. Figure 4.7 depicts the number of total CPU cycles for a single filter transformation of lengths N and several block lengths B. Here, the transform sizes have been optimized for the stream processing, as described in section 4.1.4. An opposite effect can be observed for the filter transformations: Larger block lengths require slightly larger transforms (Eq. 4.1) to be computed and can moderately increase the computational costs.

Finally, the computational costs of the methods are compared to direct time-domain filtering. An N-tap direct-form FIR filter requires N multiplies and $N - 1$ additions (Eq. 2.10) per output sample, regardless of the block length B. Therefore, the latter was assumed to have a computation time per output sample $T_{\text{TDL}}(N) = T_{\text{MUL}}(N) + T_{\text{ADD}}(N)$. For long filters, this simple cost model is representative. However, it might not be accurate for short filters $N < 64$, as advanced implementation techniques (e.g. efficient vectorization and code templates) are neglected. For block-based FFT-convolution the costs depend on the block length B. The computational savings were assessed for the typical block lengths listed above. Figure 4.8 shows the prognosed speedups of FFT-based running convolution over the time-domain filtering. Even for short block lengths B the reduction in complexity is immense and in the range $7 - 11\times$.

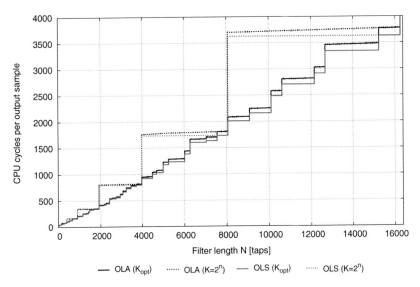

Figure 4.3.: FFT-based OLA and OLS convolution on the test system: Computational costs for streaming filtering (block length $B = 128$, solid lines correspond to convolutions using optimally sized FFT transforms, dotted lines to those with powers-of-two transforms only)

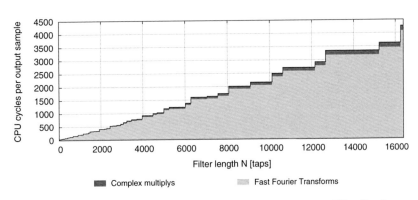

Figure 4.4.: FFT-based OLS convolution on the test system: Distribution of streaming filtering costs (block length $B = 128$)

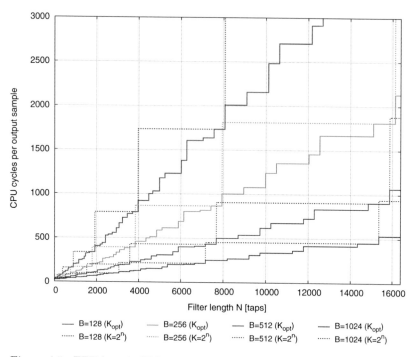

Figure 4.5.: FFT-based OLS convolution on the test system: Costs for streaming filtering (solid lines correspond to convolutions using optimally sized FFT transforms, dotted lines to those with powers-of-two transforms only)

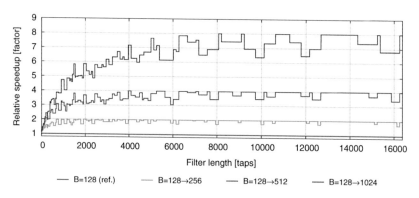

Figure 4.6.: FFT-based OLS convolution on the test system: Relative costs for streaming filtering

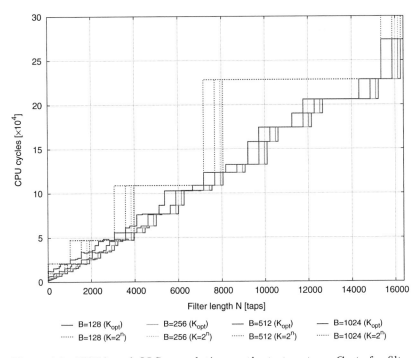

Figure 4.7.: FFT-based OLS convolution on the test system: Costs for filter transformation (solid lines correspond to convolutions using optimally sized FFT transforms, dotted lines to those with powers-of-two transforms only)

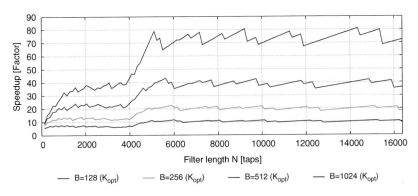

Figure 4.8.: FFT-based OLS convolution on the test system: Speedup over direct time-domain FIR filtering

4.1.6. Conclusions

- **Performance**: Unlike for time-domain filtering, the computational costs of the FFT-based OLA and OLS running convolution algorithm depends on the block length B. For typical parameters of real-time applications, the FFT-based running convolution algorithms clearly outperform simple time-domain FIR filters by several magnitudes. The exact break-even points between time-domain and frequency-domain processing require a careful inspection of mature implementations of both techniques. Asymptotically however, time-domain filtering is faster, due to a lower complexity class.

- **Distribution of costs**: In FFT-based OLA/OLS convolution algorithms, the largest part of the runtime (typically more than 90%) is spent on computing fast Fourier transforms.

- **Transform sizes**: By limiting transform sizes K on powers-of-two, filter lengths N can be found for which the OLA/OLS algorithms' performance is approximately halved (in particular when $B + N - 1$ just exceeds a power-of-two). This holds for the stream processing and the filter transformations. Hence, OLA/OLS convolution (without filter partitioning) can be significantly accelerated by also considering non-power-of-two FFT sizes. The optimal transform size can be found by inspection.

- **Cost vs. latency**: The computational costs of stream processing in OLA/OLS convolution are related almost anti-proportionally to the block length. For filter lengths $N < 6000$ taps, these ratios become more dense (shorter block lengths are comparably less computationally expensive).

  In contrast, filter transformations become computationally cheaper for smaller block lengths. Given a filter lengths $N < 6000$ the dependency on the block length can be significant. Towards longer filters $N > 6000$, the influence alleviates.

4.2. Filters with multiple inputs and outputs

So far, linear time-invariant (LTI) systems with a single input and a single output (SISO) have been regarded. Transform-based convolution can be easily extended to implement single-input multiple-output (SIMO), multiple-input single-output (MISO) and multiple-input multiple-output (MIMO) real-time FIR filters as well. A multiple-input multiple-output (MIMO) real-time FIR filter is considered in the following as a system with P inputs and Q outputs. Each of the inputs can be connected with an output over an intermediary FIR filter. $h_{i \to j}(n)$ denotes the finite impulse response of length $N_{i \to j}$ connecting the ith input and j^{th} output. It is assumed, that all input and output streams share a common sampling rate f_S and streaming block length B. For real-time audio processing these assumptions hold, as the streaming is typically driven by a single audio device. In the following it is regarded how these types of FIR filters can be realized computationally efficient using FFT-based convolution.

The principle idea is to transform input and output blocks only once. Then, only the spectral convolutions have to be computed for each filter path between an input and an output. An input spectrum can be addressed by multiple outputs. Several filter paths joining in an output are realized by summation of the computed DFT spectra. As the transforms consume the most computation (see Fig. 4.4), significant computational savings can be achieved. A requirement is that all input and output blocks and filters are transformed, using a common transform size K. As the filters $h_{i \to j}(n)$ can vary in their lengths $N_{i \to j}$, the condition to prevent time-aliasing in the outputs (Eq. 4.1) must hold for *all* of them

$$K \geq B + \left(\max_{i,\,j} N_{i \to j} \right) - 1 \qquad (4.10)$$

Two cases are of particular interest: Firstly, the case that a single input channel is filtered with two separate impulse responses. Secondly, the reversed case, when two separate input signals are filtered with the same filter impulse response. In the following it is reviewed, how complex-valued transforms allow accelerating the processing in these cases. These results can also be applied to filter with more than two inputs or outputs.

4.2.1. Dual channel convolutions

Zölzer [130] presents how the convolution of two real-valued signals can be accomplished using an equivalent complex-valued formulation. Let $x_0(n)$ and $x_1(n)$ be two real-valued length-K signals. They are combined into a complex-valued signal $x(n)$, where $x_0(n)$ forms the real part of $x(n)$ and $x_1(n)$ the imaginary

$$x(n) = x_0(n) + ix_1(n) \qquad \text{with} \qquad \begin{aligned} x_0(n) &= \Re\{\, x(n)\,\} \\ x_1(n) &= \Im\{\, x(n)\,\} \end{aligned} \qquad (4.11)$$

The K-point DFT transforms of these signals hold

$$\begin{aligned} X_0(k) &= \mathcal{DFT}_{(K)}\{\, x_0(n)\,\} \\ X_1(k) &= \mathcal{DFT}_{(K)}\{\, x_1(n)\,\} \end{aligned} \quad \Rightarrow \quad \begin{aligned} X(k) &= \mathcal{DFT}_{(K)}\{\, x(n)\,\} \\ &= X_0(k) + iX_1(k) \end{aligned} \qquad (4.12)$$

From the combined DFT spectrum $X(k)$ the DFT spectra $X_0(k), X_1(k)$ of the real and imaginary parts can be recovered, using the following relations (for details refer to [130])

$$X_0(k) = \frac{1}{2}[X(k) + \overline{X(K-k)}\,] \qquad X_1(k) = -\frac{i}{2}[X(k) - \overline{X(K-k)}\,] \quad (4.13)$$

The relations 4.11 and 4.13 make it possible to compute two K-point real-data DFTs by using just a single complex-valued K-point FFT. Based on the combined spectrum $X(k)$, two circular convolutions $y_0(n) = x_0(n) \circledast h(n)$ and $y_1(n) = x_1(n) \circledast h(n)$ of *two* separate signals $x_0(n), x_1(n)$ with a *single* signal $h(n)$ can be realized simulataneously [130]

$$x(n) = x_0(n) + ix_1(n) \qquad\qquad\qquad (4.14)$$

$$X(k) = \mathcal{DFT}_{(K)}\{\,x(n)\,\}, \quad H(k) = \mathcal{DFT}_{(K)}\{\,h(n)\,\} \qquad (4.15)$$

$$Y(k) = X(k) \cdot H(k) \qquad\qquad\qquad (4.16)$$

$$y(k) = \mathcal{DFT}_{(K)}^{-1}\{\,Y(k)\,\} \qquad\qquad\qquad (4.17)$$

$$y_0(n) = \mathfrak{Re}\{\,y(k)\,\}, \quad y_1(n) = \mathfrak{Im}\{\,y(k)\,\} \qquad (4.18)$$

Such a processing is of particular interest for binaural synthesis, where a single monaural source signal is filtered with two head-related impulse response (HRIR) filters. Note, that for this dual channel convolution the separation in Eq. 4.13 is not required. It should be remarked here, that two independent convolutions $x_0(n) \circledast h_0(n)$, $x_1(n) \circledast h_1(n)$ cannot be realized using simple complex-multiplications as in Eq. 4.16. This can be seen from $[x_0(n) + ix_1(n)] \circledast [h_0(n) + ih_1(n)] = [x_0(n) \circledast h_0(n) - x_1(n) \circledast h_1(n)] + i[x_0(n) \circledast h_1(n) + x_1(n) \circledast h_0(n)] \neq [x_0(n) \circledast h_0(n)] + i[x_1(n) \circledast h_1(n)]$. Linear convolutions $x_0(n) * h(n), x_1(n) * h(n)$ of sequences with lengths M_0, M_1 and N, can be achieved by appropriate zero-padding, similar to the explanations in Sec. 2.3. The condition for the transform size K then reads $K \geq \max(M_0, M_1) + N - 1$ (cp. Eq. 2.18).

The costs for stream processing of the dual channel technique are derived from the operations in Eq. 4.14-4.18. Steps 4.14 and 4.18 do not require extra operations and are just interleaved memory accesses. The input transform in 4.15 is complex-valued. Step 4.16 requires N complex-valued multiplies (no symmetries). The inverse transform in 4.17 is as well complex-valued. This resulting costs are

$$T_{\text{OLS-CFFT}}^{\text{stream}}(B, N, K) := \frac{1}{B}\Big[T_{\text{FFT-C2C}}(K) + T_{\text{CMUL}}(K) + T_{\text{IFFT-C2C}}(K)\Big] \quad (4.19)$$

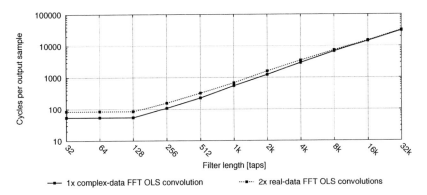

Figure 4.9.: Dual channel OLS convolution vs. two single channel OLS convolutions: Costs for stream filtering (block length $B = 128$)

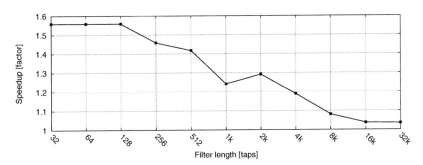

Figure 4.10.: Dual channel OLS convolution vs. two single channel OLS convolutions: Speedup for stream filtering (block length $B = 128$)

Performance

In theory, a length-N complex-data FFT is slightly more expensive than two real-data FFTs of the same length: $T_{\text{FFT-C2C}}(N) / T_{\text{FFT-R2C}}(N) > 2$. This is found by inspecting the arithmetic counts of the recent split-radix algorithms [50] for transform sizes $N > 2$. In practice however, the situation is different: the benchmarks disclosed, that a single complex-data FFT computes a little bit faster than two real-data transforms (see chapter 7). An *average* ratio $T_{\text{FFT-C2C}}(N) / T_{\text{FFT-R2C}}(N) \approx 1.79/0.97 = 1.85$ can be identified, by inspecting table 7.2. The potential of the dual channel convolution technique is examined in the following.   Figure 4.9 shows the resulting stream processing costs on the test system. The block length is $B = 128$ and power-of-two transform sizes K were selected according to Eq. 4.7. The black line corresponds to the dual channel technique using complex-valued transforms (Eq. 4.19). The dotted line represents the cost of two parallel single-channel running convolutions implemented using real-data transforms (Eq. 4.9). The speedup is plotted in figure 4.10. For short lengths $N \leq 128$ the speedup is approximately 1.56×, until for length $N \geq 16384$ it converges to the factor 1.03.

From these observations a clear conclusion can be drawn. Given that two signals need to be filtered with one impulse response or the opposite case, that one signal is filtered with two impulse responses, the dual channel technique has a definite benefit. This holds in particular for shorter filters, where the largest savings can be achieved. For very long filters the speedup is minimal. Such long filters are mostly implemented with partitioned convolution algorithms.

4.3. Filter networks

This section discusses how assemblies of FIR filters (introduced in chapter 1, section 1.3) can be realized efficiently using OLA/OLS FFT-based fast convolution. The idea is to realize both types of arrangements entirely in the frequency-domain. This is motivated by the following potential savings: In parallel arrangements of filters, the number of transforms can be reduced by reusing the input spectrum and by summing up the output spectra. In sequences of filters, the inverse transform of the last filter is followed by a forward transform of the next one. Both principles can not be used without precautions for the transform size. The necessary conditions are derived and reviewed in the following.

The partitioned input signal $x(n)$ is filtered with M different filters which are either aligned sequentially or in parallel. The M filters are assumed having finite-length impulse responses $h^{(0)}(n), \ldots, h^{(M-1)}(n)$ of lengths $N_0, \ldots, N_{M-1}$. The OLA/OLS convolution is again performed using fast Fourier transforms of size K. All impulse responses remain unpartitioned and are processed as a whole. Afterwards, the computational performance is examined in examples.

Sequential filters

Given a sequential arrangement of the filters $h^{(0)}(n), \ldots, h^{(M-1)}(n)$ as in figure 1.1(b), the M impulse responses can be merged into a single filter $h(n)$. This is achieved by time-domain convolutions or spectrum multiplications in the frequency-domain.

$$h(n) = h^{(0)}(n) \; * \; \cdots \; * \; h^{(M-1)}(n) \; = \; \mathop{\text{\huge{*}}}_{i} h^{(i)}(n) \tag{4.20}$$

$$H(k) = H^{(0)}(k) \; \cdots \cdot H^{(M-1)}(k) = \prod_{i} H^{(i)}(k) \tag{4.21}$$

Following section 2.2 the merged filter $h(n)$ has an accumulated length of

$$N = (\ldots ((N_0 + N_1 - 1) + N_2 - 1) + \ldots N_{M-1} - 1)$$

$$= \left(\sum_{i} N_i \right) - M + 1 \tag{4.22}$$

An alias-free linear convolution is only guaranteed, if the transform size meets the condition

$$K \geq B + \left(\sum_{i} N_i \right) - M \tag{4.23}$$

This inequality is obtained by inserting Eq. 4.22 into Eq. 4.1. With both methods, OLA and OLS, a cascade of filters can simply be realized by multiplying the input spectrum $X(k)$ consecutively with all filter spectra $H^{(i)}(k)$

$$Y(k) = X(k) \cdot H^{(0)}(k) \cdot \; \cdots \; \cdot H^{(M-1)}(k) \tag{4.24}$$

$$= X(k) \cdot H(k) \tag{4.25}$$

When using Eq. 4.24, the M spectrum multiplications are directly performed in the context of the audio streaming. In contrast, Eq. 4.25 first merges all filters together and then performs only a single spectrum multiplication in the context of stream processing. The number of total spectrum multiplications remains identical, but computational load is shifted away from the time-critical processing of the audio stream to the less prioritized filter update context.

Parallel filters

A parallel assembly of the M filters has the sum impulse response

$$h(n) = h^{(0)}(n) + \cdots + h^{(M-1)}(n) = \sum_i h^{(i)}(n) \qquad (4.26)$$

$$H(k) = H^{(0)}(k) + \cdots + H^{(M-1)}(k) = \sum_i H^{(i)}(k) \qquad (4.27)$$

The length N of the accumulated filter $h(n)$ is determined only by the longest impulse response in the summation

$$N = \max\{N_0, \ldots, N_{M-1}\} \qquad (4.28)$$

Hence, the condition for alias-free linear convolution of the parallel filters is

$$K \geq B + \max\{N_0, \ldots, N_{M-1}\} - 1 \qquad (4.29)$$

As before, two variants of implementation exist

$$Y(k) = (X(k) \cdot H^{(0)}(k)) + \cdots + (X(k) \cdot H^{(M-1)}(k)) \qquad (4.30)$$
$$= X(k) \cdot H(k) \qquad (4.31)$$

Eq. 4.30 is the direct implementation, requiring M spectrum multiplications and $M - 1$ spectrum additions. Eq. 4.31 uses the accumulated impulse response and requires only a single spectrum multiplication. Again, this shifts away computation from the stream processing into the filter updates. Moreover, there is also a difference in the computational effort, as spectrum multiplications are more expensive than spectrum additions.

Performance

Both types of structures can be conveniently incorporated into frequency-domain convolution, where they are mapped to multiplications or additions of filter spectra. Thereby, multiple costly forward and inverse transforms can be saved. Actually, only a single forward and inverse transform is required for each processed block. On the other hand, the new conditions for the transform size (Eq. 4.23 and Eq. 4.29) enforce the use of larger transforms, at least for sequential filters. An obvious conclusion on the combined approach can not be drawn here. For a sound comparison, the computational costs have been evaluated for the test system. This comparison includes an individual implementation of all filters and the presented combined frequency-domain techniques. The examination has been carried out with the OLS scheme and transform sizes have been specifically optimized (similar to Eq. 4.8). As there are lots of possible combinations of short and long filters, two classes of cases have been selected. Firstly: Arrangements of two, three and four equally-sized 128-taps filters. Secondly: Combinations of a 128-tap filter with longer filters of 256, 512 and 1024 taps.

Table 4.1 shows the results for sequential filters and table 4.2 for parallel assemblies. The observations can be summarized as follows:

- By implementing filter structures directly in the frequency-domain, the computational effort for the sheer filtering (streaming processing) can be lowered, often significantly.

- In case of sequential filters, the filter transformations become more expensive. In this respect, an individual implementation is computationally cheaper in total. Given that not all filters are time-varying, this gap increases even more.

- Sequences of filters can be significantly accelerated by implementing them in the frequency-domain, with respect to filtering only (streaming processing).

- Parallel assemblies of filters can be accelerated even more, by the order of several magnitudes.

- For most real-time applications, it will be beneficial to implement shorter sequences of filters and an arbitrary number of parallel filters in the frequency-domain. An individual inspection is advised, when filters are exchanged with very high rates and also in cases when not all filters in the structure are time-variant.

4.4. Filter exchange strategies

This section regards possibilities for implementing a time-varying FIR filtering, by exchanging the filters in a FFT-based fast convolution. The general description of the problem was given in section 1.3 in chapter 1. Issues of *parameter crossfading* [53] are not addressed here. It is remarked, that these techniques can be united with fast convolution techniques (e.g. HRTF interpolation in the DFT domain) and offer potential computational benefits. However, such combined approaches are very application-specific and hence not taken into consideration. Here, the focus lies on the efficient implementations of *output crossfading*, which is introduced below. Different implementation strategies are presented. The increase in computational costs for a time-varying filtering compared to a static filtering are evaluated.

Block length B	Filter lengths N_i	Cost stream individual [cyc/sample]	Cost stream combined [cyc/sample]	Speedup stream processing	Cost ftrans individual [cyc]	Cost ftrans combined [cyc]	Speedup filter transf.
128	[128, 128]	84.38	62.50	1.350	4869	7728	0.630
128	[128, 128, 128]	126.57	78.58	1.611	7304	15139	0.482
128	[128, 128, 128, 128]	168.76	106.92	1.578	9738	27911	0.349
128	[128, 256]	104.69	78.58	1.332	6070	9896	0.613
128	[128, 512]	149.11	120.74	1.235	8859	14518	0.610
128	[128, 1024]	243.30	205.78	1.182	14775	25434	0.581

Table 4.1.: Computational costs for frequency-domain implementations of sequential assemblies of FIR filters

Block length B	Filter lengths N_i	Cost stream individual [cyc/sample]	Cost stream combined [cyc/sample]	Speedup stream processing	Cost ftrans individual [cyc]	Cost ftrans combined [cyc]	Speedup filter transf.
128	[128, 128]	85.23	42.19	2.020	4869	2544	1.914
128	[128, 128, 128]	128.27	42.19	3.040	7304	2653	2.753
128	[128, 128, 128, 128]	171.32	42.19	4.061	9738	2762	3.526
128	[128, 256]	105.54	62.50	1.689	6070	3825	1.587
128	[128, 512]	149.96	106.92	1.403	8859	6775	1.308
128	[128, 1024]	244.16	201.11	1.214	14775	13021	1.135

Table 4.2.: Computational costs for frequency-domain implementations of parallel assemblies of FIR filters

Output crossfading

For simplicity it is assumed, that a filter exchange $h_0(n) \rightarrow h_1(n)$ happens within a single length-B block, here denoted as $x(n)$. Over the time span of the filter exchange, this signal block is filtered with both impulse responses, resulting in two intermediate output signals

$$y_0(n) = x(n) * h_0(n) \quad \text{and} \quad y_1(n) = x(n) * h_1(n) \tag{4.32}$$

In order to smooth out potential discontinuities, both intermediate output signals are crossfaded using envelopes $f_{\text{out}}(n)$ and $f_{\text{in}}(n)$ within the first $L \leq B$ points

$$y(n) = \begin{cases} y_0(n) \cdot f_{\text{out}}(n) + y_1(n) \cdot f_{\text{in}}(n) & \text{if } n < L \\ y_1(n) & \text{otherwise} \end{cases} \tag{4.33}$$

The length L of the crossfade is application specific. Typical values in binaural synthesis are between 8 and 32 samples. For time-varying filtering with room-impulse responses, L can be longer and even $L = B$. The envelopes $f_{\text{in}}(n), f_{\text{out}}(n)$ are defined for values $0 \leq n < L$. $f_{\text{out}}(n)$ is a non-increasing sequence and realizes the *fade out*. $f_{\text{in}}(n)$ is non-decreasing and used to *fade in*. It is claimed, that the envelopes preserve a constant amplitude

$$f_{\text{out}}(n) + f_{\text{in}}(n) = 1 \quad (0 \leq n < L) \tag{4.34}$$

Otherwise, the continuous output stream would be modulated, causing audible artifacts. From Eq. 4.34 it follows, that $f_{\text{in}}(n)$ is given by $1 - f_{\text{out}}(n)$ and vice versa. Two examples of envelope functions used for audio rendering are

$$\text{Linear} \qquad f_{\text{out}}(n) = 1 - \frac{n}{L} \qquad f_{\text{in}}(n) = \frac{n}{L} \tag{4.35}$$

$$\text{Cosine-square} \qquad f_{\text{out}}(n) = \cos^2\left(\frac{\pi n}{2L}\right) \qquad f_{\text{in}}(n) = \sin^2\left(\frac{\pi n}{2L}\right) \tag{4.36}$$

Both functions hold $f_{\text{out}}(0) = f_{\text{in}}(L) = 1$ and $f_{\text{out}}(L) = f_{\text{in}}(0) = 0$. Semantically, the first output sample $y(0) = y_0(0)$ is taken from exclusively and the sample $y(L) = y_1(L)$ already belongs to the destination signal (see Eq. 4.33).

4.4.1. Time-domain crossfading

The crossfade in Eq. 4.33 can be incorporated into the OLS running convolution algorithm in figure 4.2. Two separate running convolutions are computed side by side. The crossfade is realized afterwards in the time-domain. The resulting algorithm, in the following named *variant (1)*, is depicted in figure

4.11. Obviously, the transformed input spectrum can be reused, saving an unnecessary second input FFT. An additional spectral convolution and inverse transform are required, compared to the time-invariant OLS filtering algorithm in figure 4.2. The L-sample crossfade of the output samples demands $2L$ multiplies and L additions. The computational costs per filtered output sample for the complete time-varying OLS algorithm are

$$T_{\text{OLS-FFT-TV1}}^{\text{stream}}(B, N, K) = \frac{1}{B}\Big[T_{\text{FFT}}(K) + 2T_{\text{CMUL}}(K) + 2T_{\text{IFFT}}(K) + \underbrace{2T_{\text{MUL}}(L) + T_{\text{ADD}}(L)}_{\text{Length-}L\text{ crossfade}} \Big] \tag{4.37}$$

The second convolution branch can be realized using the dual channel convolution technique introduced in section 4.2.1. This leads to computational benefits in practice, but not in theory. The later theoretical analysis considers two complex-to-real IFFTs.

The costs for a time-varying filtering are mainly increased over those of a time-invariant filtering by the required second inverse transform. It can be avoided, by filtering the L samples of $y_0(n)$ using a direct running convolution in the time-domain (Eq. 2.2). This requires L multiplies and $L-1$ additions for each of the L partial output samples. In the following, this strategy is regarded as *variant (2)*. The computational costs of the approach are

$$T_{\text{OLS-FFT-TV2}}^{\text{stream}}(B, N, K) = \frac{1}{B}\Big[T_{\text{FFT}}(K) + T_{\text{CMUL}}(K) + T_{\text{IFFT}}(K) + \underbrace{L \cdot (T_{\text{MUL}}(N) + T_{\text{ADD}}(N - 1))}_{N\text{-tap time-domain filter}} + \underbrace{2T_{\text{MUL}}(L) + T_{\text{ADD}}(L)}_{\text{Length-}L\text{ crossfade}} \Big] \tag{4.38}$$

In principle, the L samples of $y_0(n)$ can be computed using any fast convolution technique from chapter 2. A separate $L \times N$ linear FFT-convolution is mostly disadvantageous. Indeed, it can be implemented with a smaller transform size $K' \geq L + N - 1$. But if $L \ll N$, K' will be marginally shorter than $K \geq B + N - 1$. These savings by the shorter FFTs will probably not compensate for necessary input K'-point FFT.

Performance

Both above mentioned strategies are evaluated for varying filter lengths N with a block length $B = 128$ and a crossfade length $L = 32$. The results are shown in table 4.3. The third column lists the computational costs per output sample for a static (time-invariant) filtering as in figure 4.2. Columns four and five consider the extended algorithm (Eq. 4.37) in figure 4.11. From column five it can be seen that time-varying filtering demands 50-60% more computation than a time-invariant filtering. The last two columns list the data for the combined approach with time-domain filtering (Eq. 4.37) Obviously, it is not beneficial to filter the $L = 32$ samples of $y_1(n)$ directly using a TDL. These disadvantages are emphasized for larger L. Summarizing, the costs for exchanging a filter increase significantly. In face of that only L additional samples must be filtered, the cost would ideally increase by a factor of $(B + L)/L$ only (neglecting the crossfade itself). In the example this is $(128 + 32)/128 = 25\%$.

4.4.2. Frequency-domain crossfading

The author proposed a method that integrates the time-domain crossfading in Eq. 4.33 directly into the DFT domain [123]. The benefit is that the second inverse transform in the algorithm in figure 4.11 is saved and computational costs for the time-varying filtering can be lowered further. However, the technique applies only to the OLS scheme and has constraints on the block length B and transform size K. Moreover, the crossfade needs to span over a complete output block $L = B$. The concept in [123] is generalized here. Its computational savings are analyzed.

Eq. 4.33 can be expressed in the DFT domain as follows

$$y(n) = y_0(n) \cdot f_{\text{out}}(n) \quad + y_1(n) \cdot f_{\text{in}}(n) \tag{4.39}$$

$$Y(k) = Y_0(k) \circledast F_{\text{out}}(k) + Y_1(k) \circledast F_{\text{in}}(k) \tag{4.40}$$

Crossfading the two signals $y_0(n), y_1(n)$ in the time-domain corresponds to the K-point circular convolution of their DFT spectra $Y_0(k), Y_1(k)$ with the DFT transforms $F_{\text{out}}(k), F_{\text{in}}(k)$ of the fade envelopes $f_{\text{out}}(n), f_{\text{in}}(n)$. The aim is to implement these two complex-valued circular convolutions efficiently. More precisely, with a number of arithmetic operations less than the according second inverse FFT in strategy 1. Therefore, the spectra $F_{\text{out}}(k)$ and $F_{\text{in}}(k)$ must be maximally sparse and contain only a few non-zero coefficients. The envelopes defined in Eq. 4.35 and 4.36 have broadband spectra. This degenerates the frequency-domain fading to a complexity of $\mathcal{O}(K^2)$ and renders the approach worthless.

The contribution of [123] is to employ the zone of discarded output samples in the transform-based OLS convolution. The principle is illustrated in figure 4.13. The samples in here can have arbitrary values, as they are discarded

4. Elementary real-time FIR filtering using FFT-based convolution

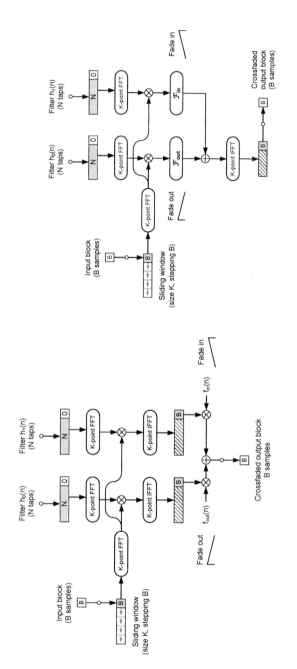

Figure 4.12.: Modified Overlap-Save FFT-convolution algorithm with frequency-domain cross-fading operators

Figure 4.11.: Overlap-Save FFT-convolution algorithm with filter exchange and time-domain crossfading

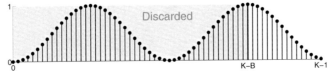

(a) Fade out within range of saved output samples

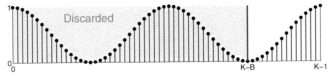

(b) Fade in within range of saved output samples

Figure 4.13.: Periodically extended fade functions for $P = 2$. The B rightmost points realize the fade out or fade in of the length-B output block. The other $K - B$ points fall within the OLS discard zone (marked gray).

and do not affect the convolution results. This allows extending the fading functions into a periodic form, which have very few DFT coefficients. For the sake of simplicity, in the following the OLS technique is modified so that the valid output samples are found at the beginning of the output buffer. This is achieved by cyclically shifting the zero-padded filter before its transformation. Particularly suited are the cosine-square envelopes in Eq. 4.36. They are now extended over the value range $0 \leq n < K$

$$\widetilde{f}_{\text{out}}(n) = \sin^2\left(\frac{\pi n P}{K}\right) = \frac{1}{2} - \frac{1}{2}\cos\left(\frac{2\pi n P}{K}\right) \tag{4.41}$$

$$\widetilde{f}_{\text{in}}(n) = \cos^2\left(\frac{\pi n P}{K}\right) = \frac{1}{2} + \frac{1}{2}\cos\left(\frac{2\pi n P}{K}\right) \tag{4.42}$$

The integer $P \in \mathbb{N}$ determines the number of covered sinusoidal periods in the interval $0 < n < K$. A strict necessity is that the right-most B samples coincide with a half of a sinusoidal period of the envelopes (cp. Fig. 4.13). This requires the transform size K to be an integer multiple of double the block length B

$$2B \mid K \quad \Rightarrow \quad P = \frac{K}{2B} \in \mathbb{N} \tag{4.43}$$

As stated above, the technique requires that the crossfade is performed over a entire length-B output block ($L = B$). In this case $\widetilde{f}_{\text{out}}(n)$ and $\widetilde{f}_{\text{in}}(n)$ cover $P = K/2B$ one or several complete periods of the cosine within the range

$0 \le n < K$. Then, the envelopes in Eq. 4.36 are obtained by extracting the B right-most points

$$f_{\text{out}}(n) = \mathcal{R}_B\{ \tilde{f}_{\text{out}}(n+K-B) \} \tag{4.44}$$

$$f_{\text{in}}(n) = \mathcal{R}_B\{ \tilde{f}_{\text{in}}(n+K-B) \} \tag{4.45}$$

The K-point DFT spectra of $\tilde{f}_{\text{out}}(n)$ and $\tilde{f}_{\text{in}}(n)$ are simple and elegant

$$\tilde{F}_{\text{out}}(n) = \mathcal{DFT}_{(N)}\{ \tilde{f}_{\text{out}}(n) \}$$

$$= -\frac{K}{4}\delta\langle k + P\rangle_K + \frac{K}{2}\delta\langle k\rangle_K - \frac{K}{4}\delta\langle k - P\rangle_K \tag{4.46}$$

$$\tilde{F}_{\text{in}}(n) = \mathcal{DFT}_{(N)}\{ \tilde{f}_{\text{in}}(n) \}$$

$$= \frac{K}{4}\delta\langle k + P\rangle_K + \frac{K}{2}\delta\langle k\rangle_K + \frac{K}{4}\delta\langle k - P\rangle_K \tag{4.47}$$

They contain only three non-zero coefficients, which are all real-valued. The DFT spectra of the continued envelopes for the case $K = 4B$, $P = 2$ (figure 4.13) are

$$\tilde{F}_{\text{out,in}}(k) = K \begin{bmatrix} \frac{1}{2} & 0 & \mp\frac{1}{4} & 0 & \dots & 0 & \mp\frac{1}{4} & 0 \end{bmatrix} \tag{4.48}$$

Inserting Eq. 4.46 and Eq. 4.47 into the K-point circular convolutions in Eq. 4.40 results in

$$Y(k) = \underbrace{Y_0(k) \circledast \tilde{F}_{\text{out}}(k)}_{=\mathcal{F}_{\text{out}}\{ Y_0(k) \}} + \underbrace{Y_1(k) \circledast \tilde{F}_{\text{in}}(k)}_{=\mathcal{F}_{\text{in}}\{ Y_1(k) \}} \tag{4.49}$$

$$= -\frac{K}{4}Y_0\langle k + P\rangle_K + \frac{K}{2}Y_0\langle k\rangle_K - \frac{K}{4}Y_0\langle k - P\rangle_K +$$

$$\frac{K}{4}Y_1\langle k + P\rangle_K + \frac{K}{2}Y_1\langle k\rangle_K + \frac{K}{4}Y_1\langle k - P\rangle_K$$

$$= \frac{K}{2}\Bigg[Y_0\langle k\rangle_K + Y_1\langle k\rangle_K +$$

$$\frac{1}{2}[Y_1\langle k + P\rangle_K - Y_0\langle k + P\rangle_K +$$

$$Y_1\langle k - P\rangle_K - Y_0\langle k - P\rangle_K] \Bigg] \tag{4.50}$$

The frequency-domain fading can be interpreted as operators $\mathcal{F}_{\text{out,in}}$ working on DFT spectra (cp. figure 4.12). The arithmetic scheme is depicted in figure 4.14 for the example $K = 4B$. Eq. 4.50 requires twelve arithmetic operations per DFT coefficient: three complex-valued additions, two complex-valued subtractions and one complex-by-real multiplication. The constant $K/2$ can

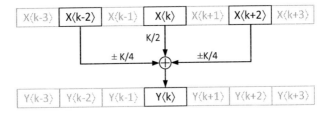

Figure 4.14.: Computation scheme of DFT domain fade operators $\mathcal{F}_{\text{out,in}}$ for the case that $K = 4B$

for instance be hidden in the filter spectrum. The arithmetic complexity of the crossfading implemented in the frequency-domain is

$$T_{\text{FD-CROSSFADE}}(K) = 3T_{\text{ADD}}(2K) + 2T_{\text{SUB}}(2K) + T_{\text{MUL}}(2K) = 12K \quad (4.51)$$

Note, that for complex-conjugate symmetric DFT spectra only $C = \lceil (K + 1)/2 \rceil$ coefficients must be processed. Figure 4.12 illustrates the final algorithm. After both fade operators have been applied to the intermediate DFT spectra, the results are summed up in the DFT domain and transformed back in the time-domain using a single IFFT. The total computational costs of the algorithm with frequency-domain crossfading are

$$T_{\text{OLS-FFT-TV3}}^{\text{stream}}(B, N, K) = \frac{1}{B} \Big[T_{\text{FFT}}(K) + 2T_{\text{CMUL}}(K) + \\ T_{\text{FD-CROSSFADE}}(K) + T_{\text{IFFT}}(K) \Big] \quad (4.52)$$

Table 4.4 lists the performance data of the proposed time-varying filtering method, which is labeled as variant (3). A comparison with table 4.3 reveals the computational benefit of the method. Here, the increase in costs for a time-varying filtering range only between 17-34%. This is a strong improvement over the time-domain crossfading algorithms presented before, which were 50-60% more expensive. The last two columns in table 4.4 show, that the frequency-domain crossfade operation in Eq. 4.50 is less expensive than an inverse transform. An advantage of the presented technique is, that it is conceptually simple and it can nicely be integrated into time-invariant convolution algorithms (cp. figures 4.2 and 4.12). However, it is not as flexible as time-domain approaches, which allow arbitrary crossfade lengths L. If this is eventually a disadvantage in practical applications (e.g. audio rendering, spatial sound reproduction) remains a topic for future studies.

Filter length N	Transform Size K	Static filtering cost stream [ops/sample]	Time-varying filtering (1) cost stream [ops/sample]	cost increase	Time-varying filtering (2) cost stream [ops/sample]	cost increase
128	256	53.3	83.7	57.1%	117.8	121.1%
384	512	121.6	189.2	55.6%	314.1	158.3%
896	1,024	273.5	423.0	54.7%	722.0	164.0%
1,920	2,048	607.4	935.9	54.1%	1,567.9	158.1%
3,968	4,096	1,335.6	2,052.2	53.7%	3,320.1	148.6%
8,064	8,192	2,913.1	4,466.5	53.3%	6,945.6	138.4%
16,256	16,384	6,309.8	9,657.5	53.1%	14,438.3	128.8%
32,640	32,768	13,586.9	20,765.1	52.8%	29,907.4	120.1%

Table 4.3.: Computational costs of static filtering and time-varying filtering using time-domain crossfading

Filter length N	Transform Size K	Static filtering cost stream [ops/sample]	Time-varying filtering variant (3) cost stream [ops/sample]	cost increase	IFFT [ops]	FD fading [ops]
128	256	53.3	71.4	34.1%	3,022	1,548
384	512	121.6	157.8	29.7%	7,014	3,084
896	1,024	273.5	345.6	26.4%	15,962	6,156
1,920	2,048	607.4	751.5	23.7%	35,798	12,300
3,968	4,096	1,335.6	1,623.8	21.6%	79,334	24,588
8,064	8,192	2,913.1	3,489.3	19.8%	174,150	49,164
16,256	16,384	6,309.8	7,462.0	18.3%	379,250	98,316
32,640	32,768	13,586.9	15,891.0	17.0%	820,406	196,620

Table 4.4.: Computational costs of static filtering and time-varying filtering using frequency-domain crossfading

4.5. Summary

Real-time FIR filtering can be easily implemented by combining FFT-based convolution with the Overlap-Add (OLA) or Overlap-Save (OLS) technique. The resulting algorithms are conceptually simple. The input-to-output latency can be adjusted by the block length of the audio stream. It is often misconceived, that a low latency would require short transforms, which is not true. The presented techniques prove the opposite. Real-time filtering is not hindered as long as the necessary computations can be executed within the given time budget. Also favorable is the constant distribution of the computational load. Each block of the stream is processed with the same operations.

Even these simple FFT-based real-time filtering concepts can hugely reduce the computational effort over simple time-domain filters. With respect to the number of arithmetic operations per output sample, the OLS technique is slightly more efficient. These differences become significant for long filters only. Much relevance has the choice of the transform size. In certain cases, the restriction on power-of-two transform sizes can almost double the computational costs. Other FFT sizes should be taken into account.

As the input data is typically real-valued, both presented convolution techniques are often implemented with real-data FFTs and IFFTs, which saves half the number of complex-valued multiplications. When multiple signals or filters are processed, complex-data transforms can offer advantages. This is particularly interesting for MIMO filters with multiple inputs and outputs. Both, the OLA and OLS method, allow applying partial transforms in similar ways. The filter transformation can be computed using a K-point FFT from a partial input of $N < K$ points. Following the results in section 2.5.4, this is only beneficial if $N \ll K$ is reasonably small compared to K. Usually this is not the case, which makes savings questionable (see conclusions in section 2.5.4). The processing of input blocks in the OLA method can be accelerated by computing the K-point input FFT from only B points. The output IFFT however, must be fully computed. In the OLS method, only B points of the K-point output IFFT are actually required. But here, the K-point input FFT (sliding window) must be fully computed and cannot be pruned. Summing up, the savings by partial FFTs can be expected only for long filters $N \gg B$. Accelerating the transforms does not change the size of the DFT spectra, which are still large. Far greater computational savings than for partial transforms are achieved by filter partitioning techniques (see chapter 5 and 6).

Assemblies of filters (cascades or parallel branches) can be realized directly in the frequency-domain, saving costly fast Fourier transforms. However, this might increase the necessary transform sizes. Still, there is mostly a computational benefit in this strategy. The costs of the stream filtering are usually lowered, whereas the filter transformations become slightly more expensive. For parallel FIR filters the computational savings can top several magnitudes. By merging the individual filters together, parts of the computation can also be taken away from the time-critical stream filtering context and shifted into the autonomous filter updates.

5. Uniformly partitioned convolution algorithms

This chapter deals with fast convolution algorithms in the class $(\mathrm{UP_S}, \mathrm{UP_F})$, that use a uniform partitioning of the signal and filter impulse response. By their own, these algorithms enable highly efficient implementations of short to medium size FIR filters. Moreover, they serve as fundamental building blocks in enhanced algorithms for long FIR filters, which use a non-uniform filter partitioning. In principle, the sub filters in a uniform filter partition can be implemented with any fast convolution approach (cp. chapter 2). However, especially transform-based fast convolution techniques greatly benefit from a uniform filter partitioning. Spectra of input blocks can be reused for all sub filters, strongly reducing the number of forward and inverse transforms in these algorithms. The considerations in this chapter are based on FFT-based implementations. They also apply to other transform-based approaches.

The chapter is organized as follows: Firstly, a standard algorithm for uniformly partitioned convolution is introduced, which uses a fixed transform size $K = 2B$ of twice the block length B. The properties and computational costs of this algorithm are reviewed and the savings over algorithms that do not partition the filter impulse response are analyzed. Secondly, the standard uniformly partitioned convolution algorithm is generalized, aiming the comprehensive study of the class of uniformly partitioned methods, including their limitations. The enhanced algorithm supports independent partitionings for the signal and filter, as well as uncommon transform sizes (e.g. non powers-of-two). The algorithmic framework is developed and the dependencies of the above stated parameters are examined. Finally, the potential of the method is compared to regular approaches. The remainder of the chapter reconsiders aspects of MIMO systems, assemblies of filters and time-varying filtering in conjunction with uniformly-partitioned convolution algorithms.

5.1. Motivation

Unpartitioned fast convolution techniques, which have been regarded in chapter 4, are an efficient solution for real-time FIR filtering with short filters, whose lengths N are close to the streaming block length B. Transform-based fast convolutions without filter partitions become inefficient, when the

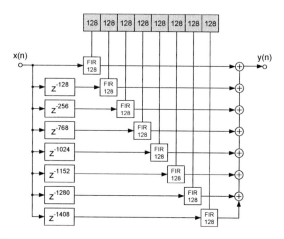

Figure 5.1.: Example of FIR filtering with a uniformly
partitioned impulse response

block length B and filter length N get out of balance $N \gg B$. The root
of this problem is found in the increasing amount of zero-padding, needed
to prolong input blocks of length B and filter of length N to the transform
size $K \geq B + N - 1$. More efficient algorithms can be derived, by splitting the large filter into several smaller ones, which on their own, are more
balanced to the block length than the original filter. From this perspective,
filter partitioning (section 3.2) is used as a real-time capable decomposition
concept. This decomposition would already be beneficial on its own. A further groundbreaking advantage of a uniform filter partitioning arises when
transform-based convolution is used as the foundation. Then the number
of costly transforms can be reduced to a minimum, resulting not only in a
massive reduction of the computational effort, but also in a lower asymptotic
runtime complexity.

The methods discussed in this chapter distinguish from previous methods.
The elementary algorithms introduced in chapter 4, considered partitioning
only in *one* operand—the signal. For offline processing, both operands can
be interchanged, due to the commutativity of convolution. Still, this would
consider partitioning exclusively in *one* operand—either signal *or* filter. In
contrast, the algorithms discussed in this chapter use a uniform partitioning
in *both* operands—signal *and* filter. Thereby, the block lengths of signal and
filter must not necessarily have the same granularity.

History

The idea of processing large signals or filters by decomposing them into smaller blocks can be found in the original publication by Stockham [105]. He referred to partitioning by the term '*sectioning*' and pointed out the computational advantages with the example of a partitioned OLA convolution. A main benefit is a strongly reduced transform size K, which becomes independent of the filter length N. The next important milestone were algorithms, which reused previously computed FFTs and accumulated the sub filter results directly in the frequency-domain. The consequence is a massive reduction in the number of transforms. Essentially, these algorithms require only a single forward and inverse transform per processed input block. The original source of this concept is hard to trace back and presumably, these 'tricks' were known and applied before. A technical paper by Kulp in 1988 [59] introduced this concept for fast real-time FIR filtering. It motivated the reuse of input spectra as well as summation in the frequency-domain. The ideas in [59] can be found in many succeeding papers on fast convolution: The works by Soo and Pang [98, 99], published by the end of the 1980s, present a similar approach in the context of adaptive filters, called multidelay block frequency domain adaptive filters (MDFs). They empirically analyze the runtime complexity and point out the computational savings over an unpartitioned processing. The realization of delays in the frequency-domain was later referred to as a frequency-domain delay-line (FDL) [37]. Torger and Farina [112] describe a PC-implementation of uniformly partitioned FFT-based OLS convolution for 3-D sound rendering and reproduction. It gives an insight into the distribution of runtime to the different operations (FFTs, spectral convolutions). Their work is closely related to the open-source convolution engine BruteFIR [111]. Opposed to the implementation by Kulp [59], it avoids a FDL by alternatively accumulating the partial convolution result in a circular frequency-domain buffer. The computational complexity remains unaffected. In a succeeding paper, Armelloni et al. [8] present a DSP implementation of the same algorithm. Wefers and Berg [118] contributed an implementation of uniformly-partitioned convolution on graphics processing units (GPUs) and measured the performance. All non-uniformly partitioned convolution techniques known to the author are assembled from uniformly-partitioned methods [39, 31, 37, 9].

The parameters of a uniformly partitioned convolution algorithm must meet specific conditions, in order to realize the processing entirely in the frequency-domain. A common choice is a sub filter size L, which is twice the block length B. Most of the above cited works found on these parameters and implement the partitioned filtering via FFT-based OLS convolution. This algorithm is referred to as standard uniformly partitioned Overlap-Save (UPOLS) in this thesis. If this strategy minimizes the computational effort of the filtering, is examined in this chapter. In [124], the author considered the use of arbitrary transform sizes (other than powers-of-two). The results indicated potential savings for long filters, but also supported the hypothesis, that the standard UPOLS method is computationally close-to-optimal for short to medium size filters. This early publication still neglected an important aspect for the frequency-domain realization, the realizability of sub filter delays. A general study of the class of uniformly-partitioned algorithm is presented in this chapter and the necessary conditions for the individual parameters are derived.

5.1.1. Uniform filter partitions

The algorithms in this chapter use a uniform partitioning of the filter impulse response. Figure 5.1 illustrates their general structure in an example. The length-N filter $h(n)$ is partitioned into sub filters $h^{(i)}(n)$ of equal length $L > 0$. The filter length N is usually given and a sub filter length L is selected as a parameter. Then the number of sub filters $P \in \mathbb{N}$ results in

$$P = \left\lceil \frac{N}{L} \right\rceil \tag{5.1}$$

An exact match $L \cdot P = N$ is not strictly necessary. The partitioning must at least cover the length of the original filter: $L \cdot P \geq N$. For the case that $L \cdot P > N$ exceeds the destination length N, the impulse response is simply zero-padded. From section 3.2 it follows, that the offset of the sub filter of index n is

$$\text{offset}(n) = n \cdot L \qquad (n = 0, \ldots, P-1) \tag{5.2}$$

In other words, the branch of the sub filter of index n in the partition must be delayed by $n \cdot L$ samples. All sub filter delays are multiples of the sub filter size.

5.2. Standard algorithm

Probably the most common uniformly-partitioned convolution method in real-time audio processing uses FFT-based Overlap-Save convolution. The standard algorithm implements a uniform filter partitioning based on FFT-convolution and combines it with the Overlap-Save method to filter consecutive audio streams. Figure 5.2 shows a block diagram of the algorithm, which is explained in the following. It filters a uniformly-partitioned input signal, provided in blocks of B samples, with a filter impulse response of N taps and outputs the results as well in length-B blocks. The two parameters (B, N) of the algorithm are the block length B and filter length N. The number of sub filters follows from Eq. 5.1. Figure 5.2 shows the variant for real-valued input data, which uses real-data transforms and makes use of symmetric DFT spectra (see section 2.5.2). This variant is examined in the following. The technique works similarly with complex-valued input data and complex transforms as reviewed in section 4.2.1.

Transform size

The standard algorithm uses a fixed transform size $K = 2B$ of twice the block length B. This allows allocating a maximum of $K-B+1 = 2B-B+1 = B+1$ filter coefficients to each DFT period, without causing time-aliasing (see section 2.5.2). However, the method only uses sub filters of length $L = B$ and leaves one sample free. This might sound negligible, but is of fundamental importance: By choosing $L = B$, the delays of the sub filter branches, as well as the accumulation of the sub filter results, can be realized entirely in the frequency-domain. The theoretical framework behind this is closely reviewed in section 5.3 As B is mostly a power of two, the FFTs of length $K = 2B$ is one as well and can be computed with maximal efficiency.

Delay-lines and accumulation in the frequency-domain

The standard technique uses the *same* granularity $B = L$ in the signal and filter partition. This allows realizing the delays of all sub filters directly in the frequency-domain. Each input block is transformed into the frequency-domain only once and the DFT spectrum is reused for all sub filters afterwards. The summation of sub filter results is realized in the frequency-domain as well, by simple addition of DFT spectra. Consequently, only a single inverse transform is necessary, in order to obtain a block of output samples. Significant computational savings are achieved, as the number of costly transforms is reduced to a minimum. Instead of computing individual FFTs and IFFTs for each sub filter, it is sufficient to compute a *single* FFT and IFFT for each input and output block.

The technique of realizing sub filter delays in the frequency-domain became known as a *frequency-domain delay-line (FDL)* [59]. An FDL is nothing else but a shift-register of spectra. For the arrival of a new length-B block, the FDL elements (spectra) are shifted by one slot. Afterwards, the new block is transformed and its spectrum is stored in the first slot of the FDL. Consequently, the FDL can only implement delays nB ($n \in \mathbb{N}_0$), which are multiples of the block length B. This affects the filter partitioning. A necessary condition to perform the processing entirely in the frequency-domain is that the sub filter length matches the block length $L = B$.

109

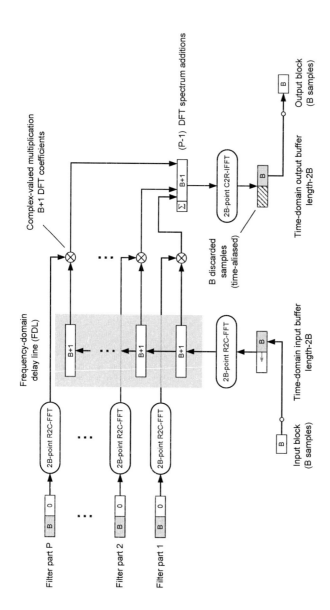

Figure 5.2.: Standard UPOLS real-data FFT-convolution algorithm with uniformly partitioned filter

The algorithm consists of two separate workflows: The partitioning and transformation of the filter impulse response and the processing of the blocks (frames) of the audio stream.

Filter processing

1. The length-N is split into $P = \lceil N/B \rceil$ length-B sub filters.

2. Each sub filter is zero-padded to length $2B$ and transformed using a $2B$-point real-to-complex FFT. Hence, each sub filter is described by $B + 1$ complex-conjugate symmetric DFT coefficients.

Stream processing

1. The input buffer acts as a $2B$-point sliding window of the input signal. With each new input block, the right half of the input buffer is shifted to the left and the new block is stored in the right half.

2. All contents (DFT spectra) in the FDL are shifted up by one slot.

3. A $2B$-point real-to-complex FFT is computed from the input buffer, resulting in $B+1$ complex-conjugate symmetric DFT coefficients. The result is stored in the first FDL slot.

4. The P sub filter spectra are pairwisely multiplied with the input spectra in the FDL. The results are accumulated in the frequency-domain.

5. Of the accumulated spectral convolutions, an $2B$-point complex-to-real IFFT is computed. From the resulting $2B$ samples, the left half is discarded and the right half is returned as the next output block.

5.2.1. Computational complexity

The computational costs for the real-valued implementation of the standard UPOLS algorithm in figure 5.2 are now derived. The filter transformation consists of P length-$2B$ real-data FFTs

$$T_{\text{UPOLS}}^{\text{ftrans}}(B, N) := P \cdot T_{\text{FFT-R2C}}(2B) = \left\lceil \frac{N}{B} \right\rceil \cdot T_{\text{FFT-R2C}}(2B) \qquad (5.3)$$

Memory transactions and zero-padding are not regarded here. In applications they can be reduced or even dropped using suitable implementations. Inserting the theoretical cost model of an FFT (introduced in section 7.4.1), with a proportionality constant k, the runtime complexity of the filter transformation results at

$$T_{\text{UPOLS}}^{\text{ftrans}}(B \text{ const.}, N) = \left\lceil \frac{N}{B} \right\rceil k2B \log_2(2B) \qquad (5.4)$$

$$= \alpha N + \mathcal{O}(1) \quad \text{(with } \alpha \text{ const.)}$$

Here, the righthand-side term depends just on the block length B and is constant. Hence, the runtime complexity of the filter transformation lies in

$$T_{\text{UPOLS}}^{\text{ftrans}}(B \text{ const.}, N) \in \mathcal{O}(N) \qquad (5.5)$$

The cost per filtered output sample (stream processing) is given by

$$T_{\text{UPOLS}}^{\text{stream}}(B, N) := \frac{1}{B} \Big[T_{\text{FFT-R2C}}(2B) + T_{\text{IFFT-C2R}}(2B) + \qquad (5.6)$$

$$T_{\text{CMUL}}(B{+}1) + (P{-}1) \cdot T_{\text{CMAC}}(B{+}1) \Big]$$

The first $B{+}1$-point CMUL operation overwrites the accumulation buffer. Hence, only the succeeding $P - 1$ spectral convolutions need to be added up and realized using CMACs. Applying inserting the theoretical costs (arithmetic operations) from section 7.4.1 yields

$$T_{\text{UPOLS}}^{\text{stream}}(B, N) = \frac{1}{B} \Big[4kB \log_2(2B) + 6(B{+}1) + 8(P{-}1)(B{+}1) \Big] \qquad (5.7)$$

$$= \frac{1}{B} \Big[4kB \log_2(2B) + 6(B{+}1) +$$

$$8 \Big(\left\lceil \frac{N}{B} \right\rceil - 1 \Big)(B{+}1) \Big] \qquad (5.8)$$

In sake of a quantitative analysis, it is assumed that $B{+}1 \approx B$ and the ceiling function is replaced by the upper bound $\lceil N/B \rceil - 1 < N/B$

$$= \frac{1}{B} \Big[4kB(1 + \log_2 B) + 6B + 8B\frac{N}{B} \Big] \qquad (5.9)$$

$$= 4k \log_2 B + 8\frac{N}{B} + 4k + 6 \qquad (5.10)$$

$$= \alpha \log B + \beta \frac{N}{B} + \mathcal{O}(1) \qquad (5.11)$$

$$\text{(with } \alpha = 4k/\log 2, \beta = 8)$$

This general asymptotic cost function holds for all fast $\mathcal{O}(N \log N)$ transforms. Note, that the logarithmic term $\alpha \log B$, which originates from the FFTs, is independent of the filter length N and therefore accounts as fix costs. Apart of an offset, the progression of the costs (Eq. 5.11) is linear, with the slope N/B. Consequently,

$$T_{\text{UPOLS}}^{\text{stream}}(B \text{ const.}, N) \in \mathcal{O}(N) \qquad (5.12)$$

Although larger block lengths B increase the costs for FFTs, they also lower the slope N/B and thereby the computational costs over the filter length N. The same complexity classes hold for uniformly partitioned Overlap-Add (UPOLA) algorithms as well. The filter transformation is identical for both. The stream processing is increased only by a linear term for adding up the overlapping samples. Where for a fixed block length B the stream processing of the unpartitioned OLA/OLS running convolutions have a time-complexity in $\mathcal{O}(N \log N)$ (Eq. 4.6), the uniform filter partition lowers the complexity class to $\mathcal{O}(N)$ (Eq. 5.12). It should be pointed out, that this is the same class as for a time-domain FIR filter (Eq. 2.10). However, the uniformly partitioned algorithms require a much lower number of arithmetic operations, as it is examined in the following.

5.2.2. Performance

Figure 5.3 shows the computational costs per filtered output for standard UPOLS algorithm (solid lines) in comparison to the unpartitioned OLS method (dotted lines) on the test system. The relative speedup of UPOLS over regular OLS is separately plotted in figure 5.4. The computational savings are significant and they increase over the filter length. Only for very short filters $N < B$ the unpartitioned method can outperform the standard UPOLS technique. This is due to the fact, that the standard UPOLS uses a fixed transform size of $K = 2B$. For very short filters $N < B$, the unpartitioned OLS can use a smaller transform size, which lowers the costs. As soon as $N \geq B$, the uniformly-partitioned algorithms were observed to be faster. For longer filters the speedup reaches the order of several magnitudes. The UPOLS method offers the largest acceleration for small block lengths B. Figure 5.7 plots the distribution of computation over fast transforms and complex-valued multiplications and additions. The results differ to regular OLS convolutions (without filter partitioning) in figure 4.4. With increasing filter lengths, the major share of the runtime is spent on spectrum multiplications and accumulations. fast Fourier transform consume a fixed number of cycles, independent of the filter length N. Figure 5.5 shows the computational costs for the filter transformation. Here it can be seen, that the multiple small FFTs required for UPOLS are more expensive, than the single large filter FFT in regular OLS.

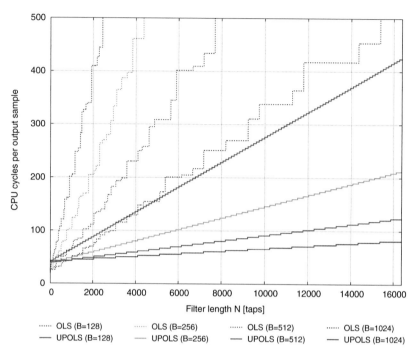

Figure 5.3.: UPOLS versus OLS convolution on the test system:
Computational costs of stream processing

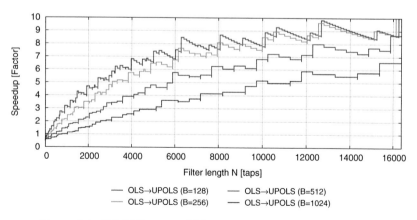

Figure 5.4.: UPOLS versus OLS convolution on the test system:
Stream processing speedup over block length

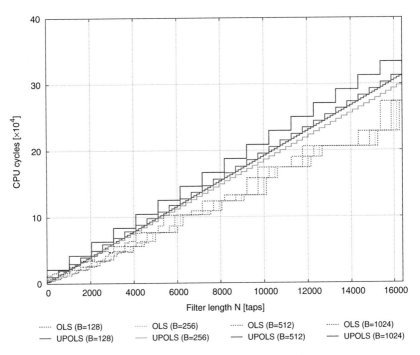

Figure 5.5.: UPOLS versus OLS convolution on the test system:
Computational costs of a filter transformation

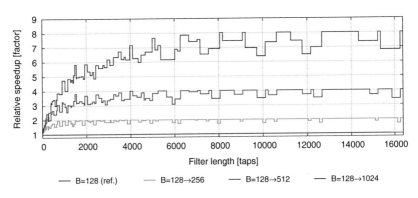

Figure 5.6.: UPOLS versus OLS convolution on the test system:
Relative computational costs of the stream processing over the
filter length for different block lengths

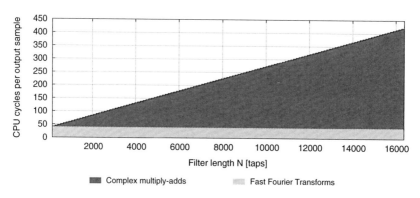

Figure 5.7.: Distribution of streaming filtering costs in UPOLS convolution
for the block length $B = 128$ on the test system.

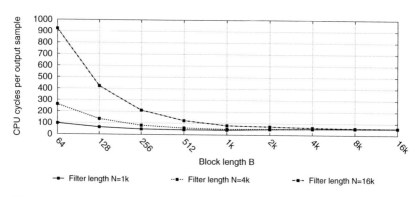

Figure 5.8.: UPOLS convolution on the test system:
Relative computational costs of the stream processing over the
block length for different filter lengths

B=128, N=384, K=512		OLS	UPOLS	OLS→UPOLS
Filter transformation	FFTs	7,014	9,066	1.29× slower
Stream processing	FFTs+IFFTs	1,4028	6,044	2.32× faster
	CMULs+CMACs	1,542	2,838	1.84× slower
	Ops/sample	121.6	69.4	1.75× faster

B=128, N=3,968, K=4,096		OLS	UPOLS	OLS→UPOLS
Filter transformation	FFTs	79,334	93,682	1.18× slower
Stream processing	FFTs+IFFTs	158,668	6,044	26.25× faster
	CMULs+CMACs	12,294	31,734	2.58× slower
	Ops/sample	1,335.6	295.1	4.53× faster

Table 5.1.: Computational complexity (number of arithmetic operations) of unpartitioned OLS and uniformly-partitioned UPOLS convolution

117

Table 5.1 gives a more detailed comparison of the costs of both approaches: filtering with an unpartitioned versus uniformly-partitioned filter. Two examples of short ($N = 384$) and medium ($N = 3968$) lengths filters are considered. The evaluation uses the theoretical costs. Therefore, combinations of filter lengths N and block lengths B are chosen, that match with power-of-two transform sizes K. For smaller filter lengths, the filter transformation of a partitioned impulse response is a little more expensive than for the unpartitioned method. However a break-even point exists and for sufficiently long filters ($N > 2^{14}$) the filter partitioning actually saves costs. With respect to the costs of pure filtering it is significantly faster in any case. For very short filters the uniform method effectively becomes an unpartitioned convolution (just a single sub filter) and the costs are equal. It achieves its computational savings mainly in the transforms. A uniform partitioning allows reducing the FFT size to the expense of more arithmetic operations needed for the spectral convolutions. The examples disclose how filter partitioning helps gaining efficiency by turning unbalanced convolution into several smaller balanced ones. This is even the case for very short filters (e.g. HRIR processing).

5.2.3. Conclusions

- A uniform filter partition allows computing FFT-based running convolutions in linear runtime $\mathcal{O}(N)$ over the filter length N. This is asymptotically faster than FFT-techniques without filter partitioning. These methods have a time complexity per filtered output samples in $\mathcal{O}(N \log N)$.

- The transform size K is small and determined by the block length B. It is however independent of the filter length N.

- Each length-B input block is represented by $K = B+1$ symmetric DFT coefficients, regardless of the filter length N. The partitioned length-N impulse response is fully described by $\left\lceil \frac{N}{B} \right\rceil (B + 1) = N + \mathcal{O}(1)$ DFT coefficients.

  In other words, a uniformly-partitioned convolution represents signal and filter by independent numbers of spectral coefficients, which are proportional to the sizes B, N of the operands. They overcome the weakness of elementary methods for the case of unbalanced convolution problems (cp. section 2.5.2).

- The computational costs for stream filtering with UPOLS convolution are several magnitudes lower than for (unpartitioned) OLS. This is reasoned by the reuse of previously computed spectra and the accumulation of sub filter outputs in the frequency-domain. The costs lower with increasing block lengths B. The speedup OLS→UPOLS has a logarithmic progression and depends also on the block length B. It is larger for shorter block lengths B, when the algorithm computes small FFTs. On the test systems, UPOLS outperforms OLS by factors of $3.5 \times -10\times$ for filter lengths $N > 6000$ and block lengths $B \in \{128, 256, 512, 1024\}$. Only for filter lengths $N < B$, the choice of non-power-of-two transform sizes makes the (unpartitioned) OLS more efficient than the standard UPOLS, which is fixed to power-of-two transform sizes. This limitation can be overcome using the generalized GUPOLS algorithm, introduced in section 5.3.

- The filter transformation of UPOLS requires more computational effort than the simple OLS approach. For unpartitioned methods it just consists of a single larger FFT. Uniformly-partitioned techniques require a multitude of small FFTs, which are in sum a little more expensive. On the test system, the computational effort was increased in the range of $10 - 50\%$ (speed up $0.6 - 0.9\times$) for filter length $N > 6000$. The relative increase in costs is larger for smaller filters and relatively independent of the block length B. However, when only few filter coefficients shall be changed, the uniformly-partitioned methods do not require a full transform to be computed, making them much more efficient.

- Compared to OLS, the UPOLS algorithms compute fewer transforms and spend more computation on spectral convolutions. Yet for short filters $N < 10B$, fast Fourier transforms still consume the dominant part of the runtime. For medium sized filters $N \sim 6000$, the computation time is roughly shared between FFTs and spectrum multiplications. Spectrum accumulations (additions) contribute only a minor part. Towards longer filters, the main part of computation shifts to spectrum multiplications.

5.3. Generalized algorithm

The study of unpartitioned OLS convolution in section 4.1 unveiled the potential of transform sizes other than powers of two for speeding up the convolution. In the literature a common approach to uniformly-partitioned convolution can be found: The starting point is typically the block length B, which is chosen with respect to the latency requirements. The impulse response is then split into sub filters of the same length $L = B$. In other words, both input signal *and* filter are uniformly partitioned with the *same* granularity. An excellent choice is a transform size K of twice the block length $K = B + L = 2B$. As B is typically a small power of two, so is K and the transforms compute highly efficient. The only two parameters of this standard algorithm are the block length B and filter length N.

Nevertheless, it is possible to use different granularities $B \neq L$ within one convolution algorithm. This section introduces a generalized uniformly-partitioned Overlap-Save (GUPOLS) convolution, for which the sub filter size L and transform size K can be varied. In line with the generalizations, the relations between the parameters (B, N, L, K) are firstly examined. With respect to a complete frequency-domain realization, only specific combinations of parameter are feasible. Several possibilities are presented and discussed. Afterwards, the generalized uniformly-partitioned Overlap-Save is presented and its computational costs are evaluated. Finally, it is examined if the generalized technique offers a benefit over the standard UPOLS method.

5.3.1. Conditions for frequency-domain processing

Uniformly-partitioned algorithms outperform methods without filter partitions by reducing the number of transforms and the transform sizes as well. Key to these improvements is, that the complete sub filter processing can be entirely realized in the frequency-domain. This includes the delays for sub filters and the summation of their outputs. The summation in the frequency-domain is trivial. It only requires that all sub filters are implemented with the same transform size. The critical point is, that the sub filter *delays* can be realized in the frequency-domain as well. An FDL can only realize delays which are multiples nB of the block length B. A general possibility to realize sample-wise delays d is given adding d zeros in the front of the sub filter impulse response. In case of transform-based convolution, it must be made sure, that the prolonged sub filters still fit the transform period and time-aliasing is avoided.

Sub filter delays

A partitioned input stream with the block length $B \in \mathbb{N}$ and a uniform filter partition consisting of sub filters of the size L are considered in the following. The block length $B \geq 1$ and sub filter length $L \geq 1$ may be freely chosen. Given a filter of length N, the impulse response is split into P sub filters with

$$P = \left\lceil \frac{N}{L} \right\rceil \tag{5.13}$$

Then the delay of the mth sub filter is mL samples ($0 \leq m < P$) (Eq. 5.2). These sub filter delays mL are now expressed by a *block-multiple delay* nB, which can be realized by an FDL, plus a *remainder delay* $d < B$, which is less than one block:

$$
\begin{aligned}
0 \cdot L &= n_0 \cdot B + d_0 &&\Rightarrow & n_0 &= 0 & d_0 &= 0 \\[2mm]
1 \cdot L &= n_1 \cdot B + d_1 &&\Rightarrow & n_1 &= \left\lfloor \frac{L}{B} \right\rfloor & d_1 &= \mathrm{mod}(L, B) \\[2mm]
2 \cdot L &= n_2 \cdot B + d_2 &&\Rightarrow & n_2 &= \left\lfloor \frac{2L}{B} \right\rfloor & d_2 &= \mathrm{mod}(2L, B) \\[2mm]
&\ \ \vdots & && &\ \ \vdots & &\ \ \vdots \\[2mm]
m \cdot L &= n_m \cdot B + d_m &&\Rightarrow & n_m &= \left\lfloor \frac{mL}{B} \right\rfloor & d_m &= \mathrm{mod}(mL, B)
\end{aligned}
\tag{5.14}
$$

Each of these lines can be interpreted as an integer division with remainder. The block-multiples n_m and remainders d_m are given by

$$n_m = \left\lfloor \frac{mL}{B} \right\rfloor \tag{5.15}$$

$$d_m = mL - n_m B = mL - \left\lfloor \frac{mL}{B} \right\rfloor B = \mathrm{mod}(mL, B) \tag{5.16}$$

$$d_m \in \{0, \ldots, B-1\} \tag{5.17}$$

The multiples n_m correspond to the block-multiple delays, which can be implemented using an FDL. In particular, n_m marks the FDL slot from where the mth sub filter takes its input spectrum. The remainder d_m corresponds to the delay portion, which cannot be realized by means of the FDL. It is here implemented by left-side padding of the sub filter impulse response. The block multiples n_m form a non-decreasing sequence $n_0 \leq n_1 \leq \ldots \leq n_m$, whereas the remainders d_m are not ordered in general.

All remainders d_m are multiples of the greatest common divisor of L and B

$$\forall m \in \mathbb{N} : \exists k \in \mathbb{N} : \quad d_m = k \cdot \gcd(L, B) \qquad (5.18)$$

This is an indirect consequence of Bézout's identity

$$mL = nB + d$$

$$\Leftrightarrow mL - nB = d$$

$$\Leftrightarrow \exists l, b \in \mathbb{N} : \underbrace{(ml - nb)}_{\in \mathbb{N}} \cdot \gcd(L, B) = d$$

$$\text{with} \quad l = \frac{L}{\gcd(L, B)}, \quad b = \frac{B}{\gcd(L, B)}, \quad L, B \in \mathbb{N}$$

Maximal remainder delays

The remainder delays are of great importance for practical implementations and especially the choice of viable transform sizes, which will be studied in the next sections. Before, the actual values of d_m and specifically the maximal possible sub filter delay $\max\{d_m\}$ are further studied here.

Given the choice of (L, B), the occurring remainder delays d_m can be computed using Eq. 5.16. This is done iteratively, for an increasing number of sub filters. The results for the first k sub filters are united into the set of sub filter remainder delays of order k, that is defined as follows

$$\mathcal{D}_k := \bigcup_{m \leq k} \{d_m\} = \{ \bmod(mL, B) \mid 0 \leq m \leq k \} \qquad (5.19)$$

Eventually, for some maximal order $k_{\max}$, this construction converges $\mathcal{D}_{k_{\max}} = \mathcal{D}_{k_{\max}-1}$ and only previously found delays will reoccur. The first sub filter never requires any delay d_0. As the remainders d_i form a periodic sequence, it follows that the first element repeated is the initial element d_0. Hence the maximal order $k_{\max} \in \mathbb{N}$ can be identified as the smallest integer $k_{\max} > 0$ that solves the congruence $k_{\max} L \equiv 0 \mod B$

$$k_{\max} = \frac{\mathrm{lcm}(L, B)}{L} = \frac{B}{\gcd(L, B)} \qquad (5.20)$$

Given a block length B and a uniform filter partition consisting of P length-L parts, the maximal remainder sub filter delay can be found as follows:

- For a number of filter parts $P < k_{\max}$, all P individual delays d_m must be computed using Eq. 5.16 and inspected

$$d_{\max} = \max\{\, d_0, d_1, \ldots, d_{P-1} \,\} \qquad (5.21)$$

- As soon as $P \geq k_{\max}$, all possible sub filter delays d_m of the form in Eq. 5.18 occurred. Hence the maximal delay is the largest multiple of $d_{\max} \cdot \gcd(L, B) < B$, given by

$$d_{\max} = B - \gcd(L, B) \qquad (5.22)$$

Examples

The established relations are now examined at two examples. The first example considers a block length $B = 32$ and sub filter size $L = 35$. The multiples n_m and remainders d_m have the form

$$n_m = \left\lfloor \frac{m \cdot 35}{32} \right\rfloor \qquad d_m = \mathrm{mod}(m \cdot 35, 32) \qquad (5.23)$$

Table 5.2 shows their values for $m = 0, \ldots, 35$. As B and L are relatively prime ($\gcd(35, 32) = 1$), the maximal order (Eq. 5.20) is $k_{\max} = B/\gcd(35, 32) = 32$. For $m = 32$ the remainder delay 0 reappears for the first time. The maximal remainder (Eq. 5.22) $d_{\max} = B - \gcd(L, B) = 32 - 1 = 31$ occurs in the 22nd sub filter ($m = 21$). The dashed lines indicate overflows modulo the block length B. The FDL slots $11, 23, 34, \ldots$ are skipped. The example also illustrates, that the actual maximal sub filter delay $d_{\max}$ depends on the number of filter parts P, unless $P > k_{\max}$. A filter of $N = 128$ taps would be partitioned into $P = \lceil 128/35 \rceil = 4$ sub filters. Then, the set of sub filter remainder delays would be $\mathcal{D}_3 = \{0, 3, 6, 9\}$ and the maximal sub filter delay $d_{\max} = 9$. The maximal delay for a 512-tap filter, that is partitioned into $P = \lceil 512/35 \rceil = 15$ parts, is $d_{\max} = 30$ (found for $i = 10$). Any filter with $P \geq 31$ sub filters incorporated the maximal remainder delay $d_{\max} = 31$.

In the second example also a block length of $B = 32$ is regarded, but the sub filter length here is $L = 36$. All delay values are printed in table 5.3. Here B and L are not relatively prime and have a common divisor $\gcd(36, 32) = 4$. As a result, there are only 7 different remainders $d_m \in \{0, 4, 8, 12, 16, 20, 24, 28\}$. The remainders d_m are cyclic with a period of $k_{\max} = 32/\gcd(36, 32) = 32/4 = 8$. The overall maximal remainder delay is $d_{\max} = 32 - \gcd(36, 32) = 28$.

5. Uniformly partitioned convolution algorithms

m	n_m	d_m	m	n_m	d_m	m	n_m	d_m	m	n_m	d_m
0	0	0	9	9	27	18	19	22	27	29	17
1	1	3	10	10	30	19	20	25	28	30	20
2	2	6	11	12	1	20	21	28	29	31	23
3	3	9	12	13	4	21	22	31	30	32	26
4	4	12	13	14	7	22	24	2	31	33	29
5	5	15	14	15	10	23	25	5	32	35	0
6	6	18	15	16	13	24	26	8	33	36	3
7	7	21	16	17	16	25	27	11	34	37	6
8	8	24	17	18	19	26	28	14	35	38	9

Table 5.2.: Sub filter delays for the example of $B = 32, L = 35$

m	n_m	d_m	m	n_m	d_m	m	n_m	d_m	m	n_m	d_m
0	0	0	9	10	4	18	20	8	27	30	12
1	1	4	10	11	8	19	21	12	28	31	16
2	2	8	11	12	12	20	22	16	29	32	20
3	3	12	12	13	16	21	23	20	30	33	24
4	4	16	13	14	20	22	24	24	31	34	28
5	5	20	14	15	24	23	25	28	32	36	0
6	6	24	15	16	28	24	27	0	33	37	4
7	7	28	16	18	0	25	28	4	34	38	8
8	9	0	17	19	4	26	29	8	35	39	12

Table 5.3.: Sub filter delays for the example of $B = 32, L = 36$

Summary

Given a block length B, a sub filter size L and a number of sub filters P, the following statements hold

1. All sub filter delays are block length-multiples $\Leftrightarrow$ $B \mid L$
   In this case all sub filter remainder delays vanish ($\forall i \colon d_i = 0$)

2. Sub filter remainder delays occur ($\exists i \colon d_i > 0$) $\Leftrightarrow$ $B \nmid L$
   They have the form $d_i \equiv k \cdot \gcd(L, B) \mod B$ ($k \in \mathbb{N}$)

3. The maximal sub filter remainder delay $d_{\max} = \max\{\, d_0, d_1, \ldots, d_{P-1} \,\}$
   1. is found by inspecting all d_i, if $P < B/\gcd(L, B)$
   2. is given by $d_{\max} = B - \gcd(L, B)$, in case that $P \geq B/\gcd(L, B)$

5.3.2. Procedure

The generalized uniformly-partitioned Overlap-Save (GUPOLS) convolution algorithm is outlined in figure 5.9. The analysis of its computational costs are found in sections 5.3.5 and 5.3.6. The considerations are carried out for the OLS variant. However, the algorithm can be realized using OLA as well. The generalized algorithm has many similarities to the standard algorithm, which was introduced in the preceding section. In the following the algorithmic procedure is explained and differences to the standard method are pointed out.

Filter processing

The main difference to the filter processing in the standard algorithm (section 5.2) is that necessary sub filter remainder delays are incorporated by left-side zero-padding of the sub filter impulse responses.

1. The length-N is split into P length-L sub filters.

2. Each length-L sub filter impulse response is padded with d_i leading zeros to incorporate the remainder delay d_i of the sub filter.

3. Additional $K - L - d_i$ zeros are padded rightmost to reach the length K.

4. The result is transformed using a K-point real-to-complex FFT. The DFT spectrum of each sub filter is described by $\lceil (K+1)/2 \rceil$ complex-conjugate symmetric DFT coefficients.

Stream processing

The processing of the audio stream differs in two significant aspects from the standard algorithm (section 5.2): First, the input and output buffers have the length K. Input and output length-B blocks are written and read right-most. Second, for the spectral convolutions each filter part spectrum is multiplied with a specific input spectrum in the FDL. The input spectrum (FDL slot) is selected by the block-multiple sub filter delay n_i. Depending on the choice of parameters (B, L, K), not every FDL slot is addressed for a spectrum multiplication.

1. The input buffer acts as a K-point sliding window of the input signal. With each new input block, its content is shifted B samples to the left and the new length-B block is stored right-most.

2. All contents (DFT spectra) in the FDL are shifted up by one slot.

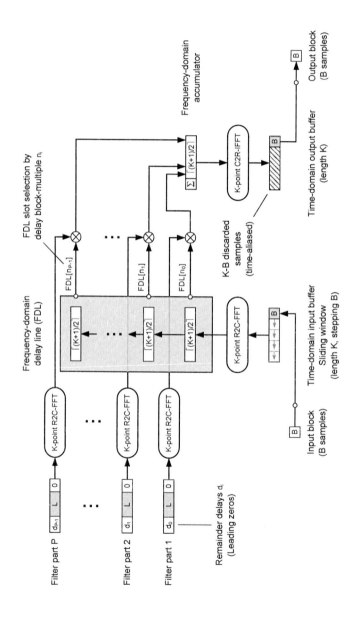

Figure 5.9.: GUPOLS FFT-convolution algorithm with adjustable transform size K and sub filter size L

3. A K-point real-to-complex FFT is computed from the input buffer, resulting in $\lceil (K+1)/2 \rceil$ complex-conjugate symmetric DFT coefficients. The result is stored in the first FDL slot.

4. All P sub filter spectra are multiplied pairwisely with corresponding input spectra in the FDL and the results are accumulated in the frequency-domain. The ith filter spectrum is multiplied with the input spectrum stored in the FDL slot n_i, where n_i is the block-multiple delay of the ith sub filter.

5. Of the accumulated spectral convolutions, a K-point complex-to-real IFFT is computed. From the resulting K samples, the $K - B$ left-most samples are discarded and B right-most samples are returned as the next output block.

5.3.3. Utilization of the transform period

The results from section 5.3.1 are now applied to obtain criteria for feasible parameters for generalized algorithms with a uniform partitioning of the signal and filter. (B, N) are in the following considered as given constraints. The sub filter size L and transform size K are free parameters.

The transform period of K points is shared by three entities, as illustrated in figure 5.10

- B points are reserved for the B-sample input blocks
- L points are allocated with filter coefficients
- D additional zeros are incorporated

The extra margin $D \geq 0$ opens the possibility to realize delays in sample granularity within the circular convolution, as motivated in section 5.3. Following the explanations in section 2.5.2, one further sample can be allocated while maintaining correct results and avoiding time-aliasing

$$K = B + L + D - 1 \qquad (K, B, L \in \mathbb{N}_0, \ D \in \mathbb{N}) \qquad (5.24)$$

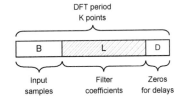

Figure 5.10.: Exploitation of the DFT transform period

Here, equality is claimed, as any oversizing accounts for D. Eq. 5.24 is fundamental for the GUPOLS algorithm and its solutions are examined in the following. The essential point is to choose (K, B, L) in a way, that the maximal occurring remainder delay d_{max} can be realized within the margin $D \geq d_{max}$

$$K \geq K_{min} = B + L + d_{max} - 1 \tag{5.25}$$

Only in this case the method can be entirely realized in the frequency-domain. All other solutions are not of interest, because they cannot be realized efficiently. Indeed, sub filter delays can also be realized in the time-domain. But in this case additional transforms are necessary.

5.3.4. Parameters

An instance of the GUPOLS algorithm has four parameters: the block length B, the filter length N, the sub filter length L and the transform size K. They can be described by a quadruple (B, N, L, K). B and N are typically preconditions. Based on their predefined values, L and K are chosen then. Feasible choices of parameters for the algorithm are examined in this section.

Full utilization

One possibility is to utilize the DFT period (Fig. 5.10) to the maximum. The K-point period just allocated with B input samples and the remaining space is used to store L filter coefficients

$$K \stackrel{!}{=} B + L - 1 \quad \Leftrightarrow \quad L = K - B + 1, \ D = 0 \tag{5.26}$$

This strategy avoids additional zero-padding ($D = 0$). Following section 5.3.1 this implies, that the only realizable sub filter delays must be multiples of the block length $L = nB$, implemented using the FDL. These conditions can now be combined into a solution for feasible transform sizes K

$$L = nB \stackrel{5.26}{=} K - B + 1 \quad \Rightarrow \quad K = B(n+1) - 1 \quad (n \geq 1) \tag{5.27}$$

For full utilization of the transform period, the transform size K must satisfy condition 5.27. This allows realizing the necessary sub filter delays directly in the frequency-domain by using an FDL. If condition 5.27 is not met, remainder delays $d_m > 0$ occur and sub filter results must first be individually transformed back into the time-domain, in order to realize the necessary delays there. This however, would make the approach very inefficient, as each sub filter would require an independent IFFT and the summation in the frequency-domain would not be possible anymore. The form of Eq. 5.27 has certain implications.

Block length B	Transform size $K = B(n+1) - 1$			
	$n = 1$	$n = 2$	$n = 3$	$n = 4$
32	63 ($3^2 \cdot 7$)	95 ($5 \cdot 19$)	127 (127)	159 ($3 \cdot 53$)
64	127 (127)	191 (191)	255 ($3 \cdot 5 \cdot 17$)	319 ($11 \cdot 29$)
128	255 ($3 \cdot 5 \cdot 17$)	383 (383)	511 ($7 \cdot 73$)	639 ($3^2 \cdot 71$)
256	511 ($7 \cdot 73$)	767 ($13 \cdot 59$)	1023 ($3 \cdot 11 \cdot 31$)	1279 (1279)
512	1023 ($3 \cdot 11 \cdot 31$)	1535 ($5 \cdot 307$)	2047 ($23 \cdot 89$)	2559 ($3 \cdot 853$)

Table 5.4.: Feasible transform sizes[4] K for full utilization of the DFT period (Eq. 5.27)

Typical block lengths in real-time audio processing are small powers of two, like $B \in \{32, 64, 128, 256, 512\}$. Table 5.4 shows possible transform sizes K for these block lengths, according to Eq. 5.27. Their prime factorization is written in brackets. All sizes are odd, most sizes contain larger prime factors and several are primes. This makes the FFTs compute rather slowly compared to highly composite sizes (e.g. powers of two, see section 7.6). In case that the application allows adapting the block length (latency), the transform size can be postulated to be a power of two and feasible block lengths are found by solving the following equation

$$K \overset{!}{=} 2^m \quad (m \in \mathbb{N}) \quad \overset{5.27}{\Rightarrow} \quad 2^m = B(n+1) - 1 \quad (n \geq 1) \tag{5.28}$$

Besides the trivial, but meaningless $B = 1$, Eq. 5.28 has also solutions of practical relevance: Some examples are ($K = 128, B = 43, L = 2 \cdot 43 = 86$), ($K = 512, B = 57, L = 8 \cdot 57 = 456$) and ($K = 4096, B = 241, L = 16 \cdot 241 = 3856$). These might be worth a consideration in hardware design, where the B is not a fixed constraint. However, for typical power-of-two block lengths, a full utilization of the transform period is for practical reasons unfavorable in uniformly-partition transform-based convolution algorithms. The required odd-length or prime-size FFTs compute comparatively slow (cp. section 7.4).

Standard algorithm

This disadvantage can be overcome by spending a margin $D = 1$ of one extra sample. The standard algorithm introduced in section 5.2 makes use of this approach. It founds on a sub filter size $L = B$ together with a transform size

---

[4]Their prime factorization is written in brackets

5. Uniformly partitioned convolution algorithms

$K = 2B$ of twice the block length B. From Eq. 5.24 it directly follows that

$$K = B + L + D - 1 \quad \Rightarrow \quad 2B = B + B + D - 1 \quad \Rightarrow \quad D = 1 \qquad (5.29)$$

Even if it is not used to implement any delay in the algorithm, this extra element is vital in order to realize the method entirely in the frequency-domain.

Power-of-two solutions

Further important solutions which prove very useful in practice, can be found by keeping the margin of a single sample. Inserting $D = 1$ into Eq. 5.27 yields solutions of Eq. 5.24

$$L = nB, \quad D = 1 \quad \overset{5.24}{\Rightarrow} \quad K = B(n+1) \quad (n \geq 1) \qquad (5.30)$$

Here the sub filter size L and the transform size K are multiples of the block length B. For standard power-of-two block lengths $B = 2^b$ ($b \in \mathbb{N}$) the resulting transform size K is highly composite and the transforms compute fast. In case that $n = 2^i - 1$ ($i \in \mathbb{N}$) it is also a power of two. Given a power-of-two transform size $B = 2^b$ and power-of-two block length $K = 2^n$ ($n \in \mathbb{N}, n > b$), the largest possible sub filter size is

$$L = K - B = 2^n - 2^b \qquad (5.31)$$

Because $2^b \mid (2^n - 2^b) = 2^b(2^{n-b} - 1) \Rightarrow B \mid L$, this method works entirely in the frequency-domain. In contrast to the standard algorithm, this approach allows using larger sub filter sizes, resulting in less sub filters, while computing power-of-two transforms only.

General solutions

So far, two types of strategies have been reviewed which found on specific assumptions: Realizations without any remainder delays ($D=0$) and solutions incorporating a single sample delay reserve ($D=1$). Now, more general conditions are derived for cases with arbitrary L and K. Thereby, the problem is addressed as follows: A sub filter size $1 \leq L \leq N$ is selected and then feasible transform sizes K for this choice are determined. Note, that for $L = N$ the algorithm results in an unpartitioned filter. A minimal transform size $K \geq K_{\min}$ exists (Cond. 5.10), which is affected by the maximal remainder delay $d_{\max}$.

For a given block length B and filter length N, the set $\mathcal{F}(B, N)$ of feasible parameter combinations (B, N, L, K) is defined as

$$\mathcal{F}(B, N) := \bigcup_{1 \leq L \leq N} \left\{ (B, N, L, K) \mid K \geq K_{\min}(B, N, L) \right\} \qquad (5.32)$$

It is constructed by iterating over all possible sub filter sizes $1 \leq L \leq N$. For each triple (B, N, L) a minimal transform size $K_{\min}$ exists. Hence, all quadruples $(B, N, L, K \geq K_{\min})$ are feasible parameter combinations for the GUPOLS algorithm.

For sufficiently long filters N, which are partitioned into at least $P \geq B/\gcd(L, B)$ sub filters (Eq. 5.20) the maximal remainder delay is known directly $d_{\max} = B - \gcd(L, B)$ by Eq. 5.22. This upper boundary is a necessary condition for any short filter. Inserting it into Eq. 5.25 yields a general condition for (L, B)

$$K \geq 2B + L - \gcd(L, B) - 1 \qquad (K, B, L \in \mathbb{N}_0) \qquad (5.33)$$

However, condition 5.33 is not sharp when $P < B/\gcd(L, B)$. Then the individual sub filters must be regarded and $d_{\max}$ must be computed using Eq. 5.21. Potentially, the minimal transform size might be smaller then stated by Cond. 5.25, because less margin D for the delays is actually needed. An example which makes this clear is $(N = 1024, B = 128, L = 136)$, where $P = \lceil 1024/136 \rceil = 8$ and $k_{\max} = 128/\gcd(128, 136) = 128/8 = 16$. The general boundary (Eq. 5.22) results at $d_{\max} = 128 - 8 = 120$, where the inspection (Eq. 5.21) of the eight sub filters unveils $d_{\max} = \{0, 8, 16, 24, 32, 40, 48, 56\} = 56$. Cond. 5.33 states that a minimal transform size $K \geq 2 \cdot 128 + 136 - 8 - 1 = 383$ is needed, but in fact $K \geq 128 + 136 + 56 - 1 = 319$ is feasible (Cond. 5.25). Only for $N \geq 16 \cdot 128 = 2048$ Cond. 5.33 is sharp.

Optimal solutions

The main interest in the GUPOLS convolution algorithm, is reasoned by its potential of lowering the computational effort over the standard UPOLS algorithm, by allowing arbitrary sub filter sizes and transform sizes. Therefore, the parameters for the GUPOLS algorithm are optimized similar to the transform size for the unpartitioned methods in section 4.1.4. Here, the optimization consists of two parameters. The sub filter length L and transform size K. The cost model is a function $T(B, N, L, K)$ of four variables. The first two variables B and N are constraints to the optimization. The optimal solution is found within the set of feasible parameters as follows

$$(L, K)_{\text{opt}} = \underset{(L, K)}{\arg\min}\{ T(B, N, L, K) \mid (B, N, L, K) \in \mathcal{F}(B, N) \} \qquad (5.34)$$

As a cost function the stream processing costs are typically used. It is not necessary to construct sets of feasible solutions in the forehand. Instead, algorithm 1 presents a strategy which computes viable parameters during the optimization. The outer loop iterates over all possible sub filter lengths L. The choice of L determines the number of sub filters P. Then the maximal sub filter remainder delay is computed. Within the range of possible transform sizes all solutions are inspected and the best is memorized.

> **Data:** Block length B, filter length N
> **Result:** Sub filter size L_{opt}, transform size K_{opt}
>
> $c_{opt} = \infty$
> for $L = 1$ to N do
> $\qquad P = \left\lceil \dfrac{N}{L} \right\rceil$
> $\qquad d_{max} = \max\{\ \mathrm{mod}(mL, B)\ \mid\ 0 \leq m < P\ \}$
> $\qquad K_{min} = B + L + d_{max} - 1$
> $\qquad K_{max} = 2^{\lceil \log_2 K_{min} \rceil}$
> $\qquad$ for $K = K_{min}$ to K_{max} do
> $\qquad\qquad c = \mathrm{cost}(B, N, L, K)$
> $\qquad\qquad$ if $c < c_{opt}$ then
> $\qquad\qquad\qquad c_{opt} = c$
> $\qquad\qquad\qquad L_{opt} = L$
> $\qquad\qquad\qquad K_{opt} = K$
> $\qquad\qquad$ end
> $\qquad$ end
> end

Algorithm 1: Brute-force optimization algorithm for the GUPOLS algorithm

5.3.5. Computational costs

The computational costs for the GUPOLS algorithm are derived similarly like in section 5.2. Again, an implementation using real-data transforms is considered. Disregarding memory copies and zero-padding, a single filter transformation consists of P length-K real-data FFTs.

$$T_{\mathrm{GUPOLS}}^{\mathrm{ftrans}}(B, N, L, K) := P \cdot T_{\mathrm{FFT\text{-}R2C}}(K) = \left\lceil \frac{N}{L} \right\rceil \cdot T_{\mathrm{FFT\text{-}R2C}}(K) \qquad (5.35)$$

The frequency-domain delay-line slot selection is just a simple lookup operation, which can be neglected. Hence, the cost for filtering a single sample of the audio stream results in

$$T_{\text{GUPOLS}}^{\text{stream}}(B, N, L, K) := \frac{1}{B}\Big[T_{\text{FFT-R2C}}(K) + T_{\text{IFFT-C2R}}(K) + \quad (5.36)$$

$$T_{\text{CMUL}}\Big(\Big\lceil \frac{K+1}{2} \Big\rceil\Big) +$$

$$\Big(\Big\lceil \frac{N}{L} \Big\rceil - 1\Big) \cdot T_{\text{CMAC}}\Big(\Big\lceil \frac{K+1}{2} \Big\rceil\Big) \Big]$$

For any sub filter $0 \le m < P$ there must be a solution to realize its required branch delay mL by an FDL delay nB $(n \in \mathbb{N})$ plus some extra delay d $(0 \le d < B)$.

$$\forall m \, \exists n, d: \quad mL = n_m B + d_m \quad (0 \le m < P, \, n_m \in \mathbb{N}, \, 0 \le d_m < B) \quad (5.37)$$

5.3.6. Performance

The performance of the GUPOLS algorithm was evaluated using the benchmark data of the test system and compared to the costs of the standard UPOLS method. The optimal choice of parameters (L, K) was selected for every problem (B, N) according to Eq. 5.34.

Figure 5.11 shows the stream processing costs over the filter length N for several block lengths B. In order to clarify these results, figure 5.12 shows the relative speedup between the UPOLS and GUPOLS method. The following observations were made:

For the case of very short filters $N < B$, the GUPOLS is superior to standard UPOLS, because it allows the use of shorter transform sizes $K < 2B$. These can lower the computational costs per output sample over UPOLS, with a fixed transform size $K = 2B$. A maximal speedup of 1.7 was found for the block length $B = 1024$. Considering the example of $B = 128$ and $N = 16$, the standard UPOLS with a transform size $K = 256$ consumes 42.2 cycles per output sample. Here, the optimized GUPOLS selects a transform size of $K = 144$, resulting in 29.6 cycles per output sample. This corresponds to a speedup of 1.42. However, it should be considered, that a 16-tap filter might be faster to implement using a matured time-domain implementation (e.g. vectorized TDL, Karatsuba technique in section 2.4.2). The savings for very short filters should therefore not be overrated.

Within a wide interval of filter lengths, the optimized GUPOLS algorithm chooses the parameters $L = B$ and $K = 2B$ of the standard UPOLS technique. This was observed for power-of-two block lengths $B \in \{128, 256, 512, 1024\}$. This proves evidence, that the standard UPOLS convolution is indeed the computationally optimal solution for these ranges of filter lengths. For the block length $B = 128$ the interval of filter lengths, where both techniques match, is $82 \leq N \leq 2944$. Until the number of $P = 23$ sub filters, $K = 2B$ remains the optimal transform size. Given longer block lengths, this span even widens. For the block length $B = 512$ the interval is $386 \leq N \leq 15872$, covering up to $P = 27$ sub filters. The range of filter lengths in which both algorithms match is surprisingly large. This has consequences for non-uniformly partitioned convolution algorithms. In these techniques the number of sub filters within segments is often even lower. In other words, the advantages of a non-uniformly partitioned convolution founding on GUPOLS instead of UPOLS are little. Only for relatively long filters $N > 25 \cdot B$, the choice of parameters differs from regular UPOLS. Here, the optimized GUPOLS technique makes use of larger sub filter lengths L, which flatten the cost curves. The break-even filter lengths N for which this takes place depends on the block lengths B. The largest relative speed up GUPOLS→UPOLS is achieved for small block lengths. In the example of $B = 128, N = 32768$ the GUPOLS is about a factor of $1.5\times$ more efficient than regular UPOLS.

Filter transformations are as well accelerated by the use of smaller transform sizes K within the same ranges of filter lengths N, as stated above. Figure 5.13 shows the costs of filter transformations for the UPOLS and GUPOLS methods. Additionally, figure 5.14 plots the relative speedups GUPOLS→UPOLS for the filter transformations. Again, the speedup depends on the block lengths and does not exceed factor two.

The most important insight is the observation, that in all optimized GUPOLS parameters the maximal remainder delay is zero. For cases where $N > B$, the transform size K was always a multiple of the block length B. Moreover, the sub filter size L was a multiple of the block length B here. This can be identified in figure 5.15 and 5.16, which show the sub filter sizes L and transform sizes K for the optimized GUPOLS methods. The concept of realizing sample-wise delays by shifting the sub filter impulse responses is eventually not beneficial. The optimal solution (B, N, L, K) selects the shortest possible transform sizes K, with respect to the efficiency of the fast Fourier transforms. Unless the filter length is very short $N \ll B$, the optimal solutions found on the strategy $K = nB$, $L = K - B$, $D = 1$ $(n \in \mathbb{N})$ (Eq. 5.30 and 5.31), which was regarded in section 5.3.4. It is concluded as a rule of thumb, that for $B < N < 20 \cdot B$ the standard UPOLS technique is optimal or near-to-optimal.

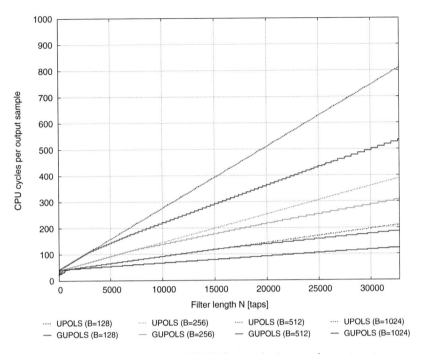

Figure 5.11.: GUPOLS versus UPOLS convolution on the test system:
Computational costs of stream processing

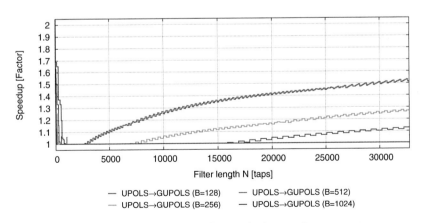

Figure 5.12.: GUPOLS versus UPOLS convolution on the test system:
Stream processing speedup over block length

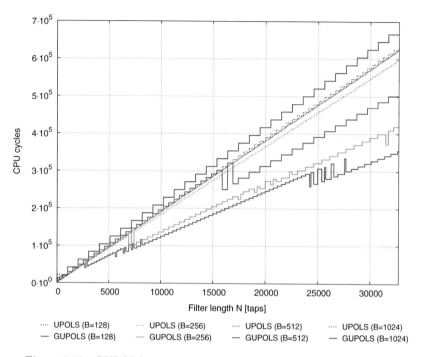

Figure 5.13.: GUPOLS versus UPOLS convolution on the test system:
Computational costs of a filter transformation

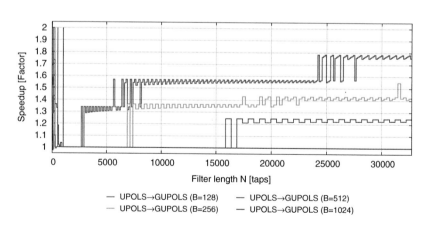

Figure 5.14.: GUPOLS versus UPOLS convolution on the test system:
Speedup of filter transformation over block length

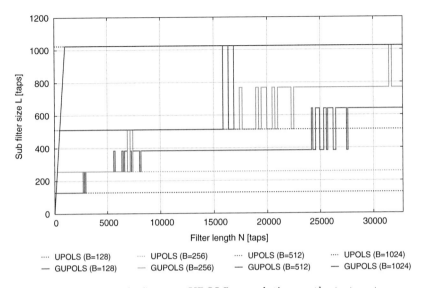

Figure 5.15.: GUPOLS versus UPOLS convolution on the test system:
Sub filter sizes over block length

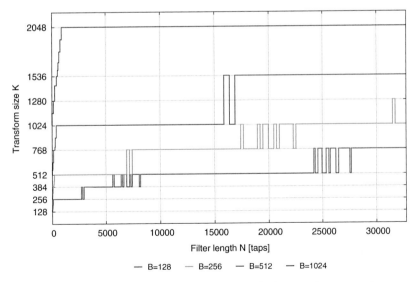

Figure 5.16.: GUPOLS versus UPOLS convolution on the test system:
Transform sizes over block length

5.3.7. Conclusions

- This section introduced a generalization of uniformly-partitioned Overlap-Save convolution, called GUPOLS. The proposed method includes both prior algorithms: (Unpartitioned) OLS and the standard UPOLS technique. Given an optimal choice of parameters, it is never slower than any of both methods.

- Optimized GUPOLS can outperform standard UPOLS with respect to the computational costs for sheer filtering (stream processing) in two cases: very short filters $N < B$, where it effectively becomes an unpartitioned OLS convolution. Here, the speedup can reach factor two ($N \to 1$). Nevertheless, time-domain filters or short linear convolution algorithms (e.g. Karatsuba technique) should be considered for $N \ll B$, which can be even faster. On the other hand, for sufficiently long filters, which exceed lengths of approximately $N > 20 \cdot B$, the GUPOLS technique makes use of other transform sizes $K \neq 2B$ and adapted sub filter sizes $L > B$. Here the speedup over UPOLS of up to 50% was observed ($B = 128, N \leq 32768$). Short block lengths B benefit more from the additional degrees of freedom in GUPOLS.

- For filter lengths in the interval of approximate $B < N < 20 \cdot B$, the optimized GUPOLS technique actually converges to the standard UPOLS algorithm with a fixed transform size $K = 2B$ and fixed sub filter size $L = B$. Consequently, for most filter lengths N of practical relevance, the standard method marks the computationally optimal solution. This holds in particular for its applications in non-uniformly partitioned convolution algorithms, which are discussed in chapter 6.

- Surprisingly, optimized GUPOLS turned out to completely avoid the use of sub filter remainder delays. For $N < 32768$ and $B \in \{128, 256, 512, 1024\}$ not a single counterexample was found.

- The set of optimal transform sizes and sub filter sizes is very limited. Unless $N < B$, both are typically powers of two or small multiples of these.

- As a rule of thumb, the transform sizes should be chosen $K = nB$, a multiple of the block length B. As the latter is usually a power-of-two, the multiple $K = nB$ computes rather efficient (e.g. $640 = 10 \cdot 64$). The sub filter length L should be chosen to fill the remaining space $L = K - B$.

5.4. Multi-dimensional convolutions

Uniformly-partitioned convolution can be regarded from yet another per-spective. Technically, the input signal $x(n)$ is partitioned into length-B blocks $x_i(n)$ and the filter $h(n)$ into length-L sub filters with impulse responses $h_i(n)$. This partitioning can be interpreted as a special case of multi-dimensional index mapping (cp. section 2.6.1) applied to linear convolution [47]. From this interpretation originate further advances, which are briefly outlined in the following.

The one-dimensional entities $s(n), h(n)$ (with a single index n) are mapped to a two-dimensional representation $x_i(n), h_i(n)$ indexed by two variables i, n. For the sake of simplicity, the considerations here are based on running convolutions using the standard UPOLS method ($L = B, K = 2B$). Burrus [20] and Hurchalla [47] provide a more general derivation of the technique, founding on finite-length operands. A central aspect is, that in the uniformly-partitioned frequency-domain convolution algorithm, each of these partitioned entities is represented by an individual spectrum

$$x_i(n) \rightarrow X_i(k), \ h_i(n) \rightarrow H_i(k) \quad \text{and} \quad Y_i(k) \rightarrow y_i(n) \tag{5.38}$$

In other words, there is a one-to-one association between blocks respectively sub filters and spectra. The definition of the spectra $X_i(k), H_i(k)$ and extraction of the output block $y_i(n)$ depends on the scheme used (OLA or OLS). The spectral processing in uniformly-partitioned convolution algorithms consists of spectral convolutions and accumulations (see figure 5.2 and 5.9). Mathematically, this can than be formalized as follows

$$Y_i(k) = \sum_{n=0}^{P-1} X_{i-n}(k) \cdot H_n(k) \tag{5.39}$$

Obviously, Eq. 5.39 marks a 1-D linear convolution, formulated on complete spectra instead of samples. Each spectral product ($\cdot$) by itself corresponds to a circular convolution. In this respect, Eq. 5.39 effectively realizes a two-dimensional convolution. Or to put it in other words, a convolution algorithm with uniform partitioning of both operands can be interpreted as a two-dimensional convolution. The decomposition of circular convolutions [20] in section 2.6.1 is a similar method. Here however, the mapping requires to be circular in *both* dimensions, which lead to incompatibility with real-time processing (details see section 2.6.1). Eq. 5.39 however, is a mixture of a linear convolution (sum) and circular convolution (spectra product). Hurchalla [47] outlined potential of this insight, which lies in the fact, that these two-dimensional convolutions can be implemented with fast convolutions along *both* dimensions. The previously reviewed algorithms (standard UPOLS and GUPOLS), evaluate Eq. 5.39 naively (direct convolution). As

P is typically short, Eq. 5.39 can be implemented using a fast short linear convolution algorithm. This can further reduce the algorithmic complexity of uniformly-partitioned techniques. However, not any short convolution algorithm is actually suitable. Hurchalla [47] reviews suitable candidates of short linear convolution algorithms and verified, that the nested Karatsuba technique (see section 2.4.2) conforms with real-time processing. The magnitude in computational savings has not been examined in detail and remains an open research topic.

5.5. Filter with multiple inputs or outputs

The presented uniformly-partitioned convolution algorithms can be extended to multiple input and outputs, following the considerations in section 4.2. Applications are found in MIMO FIR filters with medium size filters (e.g. 1k-4k taps). Typical examples are multi-channel reproduction systems, like crosstalk cancellation [62] and ambiophonics [112]. The results in section 5.2.2 showed, that a fast convolution with a uniform filter partitioning is superior over simple concepts without filter partitioning. For these lengths of filters the computational savings reach several magnitudes.

An essential criterion for MIMO UPOLS filters is, that all inputs and outputs use a common transform size K. In this case all intermediary filters $h_{i \to j}(n)$, connecting inputs and outputs, are compatible and share common transforms. For the standard UPOLS algorithm, the transform size remains $K = 2B$. The lengths of the impulse responses $h_{i \to j}(n)$ only affects the number of length-B sub filters. Each input is associated with its own FDL. The number of elements in each of these FDLs is determined by the maximal number of sub filters for all outgoing filters $h_{i \to j}(n)$. Each output is a point of superposition, followed by a single inverse transform. In total, a M-input N-output filter can be realized using M forward FFTs and N inverse FFTs. These can be real- or complex-valued, as discussed in section 5.2.

The dual-channel technique introduced in section 4.2.1 applies for uniformly-partitioned methods as well. Two independent channels or two independent filters can be processed by using complex-valued FFTs. The relative speedup of this strategy is less pronounced compared to the prior algorithms which do not partition the filter impulse responses. This is the case, because relatively little time is spent on computing fast Fourier transforms in uniformly-partitioned algorithms (cp. figure 4.4 and figure 5.7). An explicit evaluation of the computational benefits of MIMO UPOLS filters over single-channel UPOLS convolution is not conducted here. The savings of a combined implementation are less distinct for the above mentioned reasons.

5.6. Filter networks

Uniformly partitioned frequency-domain filters can be connected and arranged in sequential or parallel structures (see figure 1.1(b) and 1.1(d)) similar to (unpartitioned) OLA and OLS running convolutions, as regarded in section 4.3. Again, it is a possibility to merge the impulse responses and convolve the input stream with a single cumulative filter. Like discussed before, this can be disadvantageous in the case of time-varying filtering with high update rates. Particularly, when different filters are updated asynchronously, it might be computationally cheaper to implement the filters separately. Nevertheless, there is still room for improvements. Joyo and Moschytz [51] discussed the general possibilities to realize assemblies of uniformly partitioned filters directly in the frequency domain. The motivation is to save costly forward and inverse transforms once more. Figure 5.7 illustrates, that FFTs are not as dominant as in unpartitioned filtering concepts. Thus the potential savings are less articulated for assemblies of uniformly-partitioned filters.

The uniformly-partitioned convolution algorithms introduced before is a direct application of the theory developed in section 5.6. UPOLA and UPOLS are parallel assemblies of OLA respectively OLS filters. A parallel arrangement of multiple uniformly partitioned convolutions can be seen as a simple extension to the regular UPOLA and UPOLS algorithms. In the following it is assumed, that all filters in the assembly use a common block length B and a standard transform size $K = 2B$. The individual filters can be either UPOLA or UPOLS convolutions, but they can not not be mixed. Then input blocks are identical to all parallel filters and can thus be reused. The same accords for the summation of the parallel branches which can be realized before the inverse transform. The transform sizes are not affected by the additional parallel branches. This is due to the fact that all uniform sub filters have equal lengths. Only the number of sub filters is influenced by the lengths of the individual impulse responses. The computational savings arise from the reduced number of FFTs and IFFTs. For multiple parallel short to medium size filters the advantages of a combined frequency-domain implementation can be significant.

Sequential arrangements of uniformly-partitioned filters are more complicated. Pairs of inverse-forward transforms are found in such cascades. For two reasons they cannot simply be dropped. Firstly, the overall transform size must be sufficiently large and respect Eq. 4.23. Joining $n \in \mathbb{N}$ separate UPOLA/UPOLS filters, each with a sub filter length $L = B$, requires a combined transform size of $K \geq B + nB - 1$. Multiples $K = nB$ of the block length B are a reasonable choice for the transform size (cp. section 5.3.4). As the block length is typically a power-of-two, FFTs of these sizes compute with a high efficiency (cp. section 7.6). Secondly, the input data of the subsequent convolutions must be conditioned depending on the scheme used. For the OLA scheme this requires to select only the first B samples from the $2B - 1$ point valid output samples. In the time-domain this corresponds to the multiplication with a rectangular window. Unfortunately, the corresponding DFT spectrum of this window is not sparse (cp. section 4.4) and thus does the frequency-domain implementation require a full K-point complex-valued circular convolution in $\mathcal{O}(N^2)$. The OLS requires the same complexity. Hence, it is indispensable to enlarge the transform size. There is no general finding whether an individual or combined implementation of sequential uniformly partitioned convolutions is preferable for the sake of a higher computational efficiency. This demands an individual inspection of the actual case and depends on the block lengths, filter lengths and number of cascaded filters.

5.7. Filter exchange strategies

All techniques for exchanging filters, introduced in section 4.4, can be applied to convolution algorithms with uniformly-partitioned filter impulse responses. Exchanging the filter in the elementary methods required the execution of a second spectral convolution (see figure 4.11). For the UPOLS technique this results into a second convolution branch, where all sub filter convolutions are executed with a second filter and accumulated into a second frequency-domain accumulation buffer. For the time-domain crossfading technique (see section 4.4.1) a second inverse transform is required, as for the unpartitioned OLS algorithm. In UPOLS convolution, the second IFFT is shorter $(K = 2B)$ and has thus less impact on the increase in computation. In contrast, the spectral convolutions consume a larger share of the computation and the time-varying filtering doubles their effort. The frequency-domain implementation of crossfading presented in section 4.4.2 can be seamlessly integrated into the standard UPOLS convolution algorithm. The crossfading operation is executed on the two frequency-domain accumulators of the two spectral convolution branches. Like before, the technique saves the second inverse transforms. However, as the impact of this is comparably little, the computational savings of the frequency-domain technique will be not as emphasized as for unpartitioned filters.

Table 5.5 gives an overview about the computational costs of different variants of the UPOLS algorithm. For different combinations of block length B and filter length N, it lists the stream processing effort in cycles per output sample. The performance data was computed using the benchmark profile of the test system. By comparing table 5.5 with tables 4.3 and 4.4 for the elementary methods the following observations can be summarized: The relative increase in computational effort, when switching from a time-invariant to a time-varying filter is larger for the UPOLS convolution. This holds for both methods, the time-domain (TD) crossfading and the frequency-domain (FD) crossfading. The reason therefore can be found in the fact, that the spectral convolutions make up for a comparably large share of the runtime in UPOLS. In contrast, they consume only a fraction of the computation in the regular OLS algorithms. However, it is important to keep in mind, that the filtering itself is by magnitudes cheaper than with the elementary techniques, which do not facilitate a filter partitioning.

Table 5.5 also lists convolutions with large block lengths B, which are not relevant for real-time UPOLS convolution, as the block lengths result in unacceptable latencies (e.g. $B = 8192$). Nevertheless, these cases occur in real-time partitioned convolution techniques with non-uniform filter partitions. If in contrast smaller crossfade lengths $L \ll B$ are considered, it might be worth to consider an autonomous convolution for the overlapping output samples, as discussed in section 4.4. In the face of the lengths of the sub filters (e.g. $N > 8k$) these should be implemented with an independent fast convolution. For the case $L \ll B \ll N$, these convolutions are highly unbalanced, making the computational advantages of such separate realizations questionable.

5.8. Summary

This chapter introduced transform-based running convolution algorithms that partition a length-N filter impulse response into uniform sub filters of length $L \leq N$. This strategy is advantageous for the following reasons: Firstly, the length-L sub filters on their own are generally more balanced than the original filters (typically $L \ll N$). Elementary techniques, like classical FFT-based convolution [105], represent both operands of length B and N by a common number of $K \geq B + N - 1$ frequency-domain coefficients. The uniform filter partition breaks this boundary and allows representing each operand independently, with a number of coefficients proportional to its size. Secondly, in particular the implementation of uniformly-partitioned filters with transform-based fast convolution techniques allows a dramatic reduction of the number of transforms in the algorithm. These savings are possible, when the delays of sub filter branches along with the accumulation of sub filter results can be realized entirely in the frequency-domain. In this case the runtime complexity class per output sample of these techniques is $\mathcal{O}(N)$ and thereby asymptotically lower than for the elementary transform-based running convolution techniques, which lie in $\mathcal{O}(N \log N)$ per output sample. The different complexity classes are reasoned by the fact that the number of forward and inverse transforms is fixed for each processed block and independent of the filter length N. Not only the computational complexity

Block length B	Filter length N	Static filtering cost stream [ops/sample]	Time-varying UPOLS (TD) cost stream [ops/sample]	Time-varying UPOLS (TD) cost increase	Time-varying UPOLS (FD) cost stream [ops/sample]	Time-varying UPOLS (FD) cost increase
128	128	42.2	67.9	61.0%	53.9	27.8%
128	256	45.2	74.0	63.6%	59.9	32.6%
128	512	51.2	86.0	67.9%	71.9	40.5%
128	1,024	63.2	110.0	74.0%	96.0	51.8%
128	2,048	87.2	158.1	81.2%	144.0	65.1%
128	4,096	135.3	254.2	87.9%	240.2	77.5%
1,024	1,024	42.5	66.6	56.8%	52.4	23.3%
1,024	2,048	45.1	71.8	59.3%	57.6	27.7%
1,024	4,096	50.2	82.1	63.5%	67.9	35.1%
1,024	8,192	60.5	102.7	69.7%	88.5	46.2%
1,024	16,384	81.1	143.9	77.4%	129.7	59.9%
8,192	8,192	56.7	89.0	57.1%	73.6	29.8%
8,192	16,384	60.1	95.8	59.5%	80.4	33.8%
8,192	32,768	66.8	109.4	63.6%	93.9	40.5%
8,192	65,536	80.4	136.5	69.7%	121.0	50.5%

Table 5.5.: Computational costs of static and time-varying UPOLS filtering implemented with crossfading in the time-domain (TD) and frequency-domain (FD) as presented in section 4.4.2

is asymptotically lower, but also the number of operations. Even for moderate filter lengths, the UPOLS filtering outperformed the elementary OLS running convolution by several magnitudes. And hence, the computational savings over simple time-domain filters are enormous.

The ever-important property for these computational advantages is, that the sub filter delays mL ($m \in \mathbb{N}$) can be realized in the frequency-domain using a frequency-domain delay-line (FDL). Therefore, they must be multiples nB ($n \in \mathbb{N}$) of the block length B of the audio stream. In principle both partitions, signal and filter, can be realized using different granularities $B \neq L$ (block lengths B and sub filter sizes L). However, specific conditions must be met, so that the complete algorithm can be realized in the frequency-domain.

The straight-forward standard UPOLS approach chooses the same granularity for both axis $B = L$ and realizes the convolution with a transform size of $K = 2B$. The presented generalization of the algorithm (GUPOLS) formulates the framework for arbitrary solutions. Its parameters are the block length B, filter length N, sub filter size L and transform size K. The constraints and couplings of these four variables have been examined in detail. The general solutions are more complicated and demand careful considerations. Several strategies to choose feasible parameter sets (B, N, L, K) have been reviewed, including the brute-force optimization. The motivation in this research was verification of the standard UPOLS technique. The results revealed, that for large ranges of filter lengths N, the popular UPOLS convolution technique with standard parameters $L = B$ and $K = 2B$ is indeed optimal or near-to-optimal. The potential of non-power-of-two transform sizes is much lower for these uniformly-partitioned convolution algorithm, than for simple concepts (see chapter 4).

Filters with multiple inputs and outputs (MIMO), as well as filter networks can be implemented using uniformly-partitioned convolution methods. Compared to elementary algorithms, the computational savings of a combined frequency-domain implementation are less pronounced, but still significant. Previously discussed concepts of time-varying filtering can be nicely integrated into convolution methods that make use of a uniform filter partition. It turned out, that the frequency-domain crossfading approach presented in section 4.4.2 is as well beneficial for the UPOLS technique. Like for combined implementations of filter networks, the transforms play a less dominant role as in elementary algorithms (see chapter 4). In contrast, a performant implementation of the spectral convolution is very important for uniformly-partitioned convolution.

5. Uniformly partitioned convolution algorithms

6. Non-uniformly partitioned convolution

This chapter considers running convolution algorithms which partition the filter impulse response into non-uniform sub filters of varying sizes. These algorithms constitute the class $(\mathrm{UP_S, NUP_F})$. The key principle of these techniques lies in the implementation of later sub filters in the impulse response with larger block lengths, without sacrificing the real-time properties. This approach significantly lowers the computational effort of the real-time FIR filtering once more and results in an even lower asymptotic runtime complexity. The computational savings are groundbreaking for artificial reverberation using FIR filters (convolution reverbs) and build an important foundation for FIR-based real-time auralization of reverberant spaces. Non-uniformly partitioned convolution algorithms are assembled from several uniformly partitioned convolution algorithms. Consequently, much of the theory of uniformly partitioned convolution applies to these algorithms as well. However, non-uniformly partitioned convolution techniques are more complex than uniformly-partitioned methods. This is reasoned by the fact, that many sub convolutions need to be executed in parallel in accordance with timing dependencies. The most important parameter of these algorithms is the non-uniform filter partitioning. It strongly affects the algorithms' properties, like the computational effort, the sheer realizability in real-time, the distribution of the computational load over the runtime and the filter exchanges.

This chapter is structured as follows: Firstly, non-uniformly partitioned convolution techniques are motivated based on the findings of the prior chapters. A non-uniform filter partitioning scheme is introduced and important termini are defined. State-of-the-art algorithms are revised, including their algorithmic properties and implementation strategies. The computational costs and runtime complexities are analyzed and compared to prior methods. The main part of the chapter is dedicated to non-uniform filter partitions and their algorithmic optimization. Pitfalls of the optimization in prior publications are outlined and an advanced optimization technique is presented, which allows obtaining practicable partitions. Finally, concepts like MIMO filters, filter structures and time-variant filtering are reconsidered in the light of non-uniformly partitioned convolution.

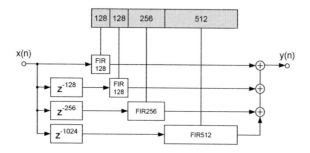

Figure 6.1.: Example of FIR filtering with a non-uniformly
partitioned impulse response

6.1. Motivation

All previously regarded frequency-domain convolution techniques revealed a dependency between their computational effort per filtered sample and the latency, in form of the block length B. For both classes of algorithms, OLA/OLS without filter partitioning, as well as UPOLA/UPOLS featuring a uniform filter partitioning, a simple and general statement can be made: Less input-to-output latency, realized by a smaller stream block length, increases the computational costs. A roughly anti-proportional relationship between the block length B and the computational costs of the filtering can be observed, given that the filters are sufficiently long (cp. figures 4.6,5.6 and 5.8).

Reconsidering the general filter partitioning scheme in figure 3.6, the following important observation can be stated: A sub filter $h_i(n)$, placed at **offset**(i) in the impulse response, contributes to the overall filtering output not before the number of **offset**(i) samples has been processed. This is easy to see by considering the corresponding TDL. Hence, the sub filter can be processed with a different block length then this of the audio stream. The input samples can be buffered and *repacked* to a larger granularity. As long as the computation of the sub filter is finished within **offset**(i) samples, the required results are ready in time and can be added into the overall output. The essential point is, that the total input-to-output latency of the filtering remains unaffected. Different regions of an impulse response can be realized using several independent partitioned convolutions that process the data in different block sizes. The initial part of the impulse response needs to be processed with the block length B of the audio stream. Towards later parts of the impulse response larger block lengths are employed. The larger the block lengths get, the more the computational effort per filtered output sample decreases.

6.2. History

The origin of non-uniformly partitioned convolution dates back to the early 1990s. The concept of exploiting a non-uniform filter partitioning for real-time FIR filtering can presumably be credited to two dutch researchers, Egelmeers and Sommen. They published the principle idea implementing sub filters of varying sizes using different block lengths in a conference paper in November 1994 [30]. This work considers implementations on DSPs. It defines non-uniform filter partitions and introduces the general structure of a non-uniformly partitioned convolution as depicted in the block diagram in figure 6.1. Sub filters are implemented using the standard FFT-based UPOLS method, reviewed in chapter 5. The computational complexity is regarded by means of real multiplications only. The authors consider the example of a medium size filter ($N = 4000$) implemented with a very small block length ($B = 4$). Five specific filter partitions are regarded. Their new approach proved to be about 16 times more efficient than standard UPOLS convolution for the stated example. Later, in December 1996, an extended work was published in an IEEE journal [31]. This extended version of the original paper also addressed aspects of adaptive filtering. Interestingly, the original manuscript of this paper was submitted by December 1993. This might be an indication, that non-uniformly partitioned convolution could be their invention.

A little bit later, Gardner presented a non-uniformly partitioned convolution method for low latency filtering, at a conference in November 1994 [38]. This work was published later in 1995 as a journal paper [39] and this particular paper became a widely-known reference for non-uniformly partitioned convolution. Gardner [38] does not include a reference to the work by Egelmeers and Sommen [38], which was published just three months earlier. Presumably, both parties invented their concepts independently. Also, the work by Gardner [38] differs from the Egelmeers' and Sommen's paper in several aspects: It also considers DSP implementations and also includes a strategy for achieving (almost) no input-to-output latency. The implementation founds on standard UPOLS as well. In the paper Gardner proposes a non-uniform filter partitioning *scheme* for *arbitrary* filter lengths. The paper also includes a false hypothesis, that the strategy of the most rapid increase in sub filter sizes results in the lowest computational costs. However, Gardner correctly points out the worst-case load-distribution of this approach, which renders it worthless in practice. A main contribution of the work is a practical non-uniform filter partitioning scheme, assembled from power-of-two sub filters, which are repeated twice before a larger sub filter size is selected. This strategy avoids unfavorable timing-dependencies and is practical.

Subsequent works considered non-uniformly partitioned convolution for realizing artifical reverberation [71, 2] (convolution reverbs) and introduced further filter partitions. Müller-Tomfelde's paper [71] from 1999 focuses on general purpose processors (no DSPs) and discusses aspects of the implementation. Later further authors addressed the rather complex real-time implementations of non-uniform convolution algorithms [76, 9]. The main challenge lies in the parallel implementation of asynchronously executed sub convolutions. Battenberg and Avizienis [9] regarded scheduling strategies on multi-core machines. Müller-Tomfelde [72] considered the problem of time-varying filtering with non-uniformly partitioned convolutions and proposed two strategies for filter exchanges in these algorithms. Recently, GPU-implementations of non-uniformly partitioned convolution were examined [82]. Primavera et al. [76] incorporated psycho-acoustic aspects to artificial reverberation implemented using partitioned convolution. The work applies a concept conceived by Yang et al. [128] to non-uniformly partitioned convolution algorithms. Computational savings are achieved by reducing the complexity of spectral convolutions by exploiting perceptual thresholds.

The choice of filter partition is not a trivial issue. A paper [37] by García in 2002 marks an important milestone. It presented an optimization technique for finding the filter partition with the least computational complexity. In this respect it is of theoretical importance, showing up the limits of the class of non-uniformly partition convolution algorithms. García formulated the construction of partitions in a state-space (weighted directed acyclic graph (DAG)), which can be effectively searched for the best path using the Viterbi algorithm within. Whilst the number of feasible filter partitions grows exponentially over the filter length, the optimization algorithm finds solutions in polynomial runtime (PTIME). Unfortunately, the results are far from practical. Wefers and Vorländer [121] pointed out the worst-case load distribution of cost-optimal filter partitions. They examined the relations between the computational costs and scheduling flexibility that a non-uniform filter partition provides. Furthermore, they presented techniques to obtain real-time capable partitions, which deliver a significant computational advantage over the method by Gardner [39]. Recently, the concept of arbitrary transform sizes and a more general framework of non-uniformly partitioned convolution has been suggested by the author in [122].

6.3. Non-uniform filter partitions

This section refines the definition of general filter partitions for the non-uniform case and introduces further important terms. In section 3.2, a filter partition was defined as a tuple $\mathcal{P} = (N_0, \ldots, N_{P-1})$ (Eq. 3.10). The elements N_i correspond to the length of the sub filters $h_i(n)$ of the partition.

Altogether they cover a total length $N = \sum N_i$ (Eq. 3.2). The sub filter lengths form a non-decreasing sequence $N_i \leq N_{i+1}$ (Eq. 3.9). Small sub filters are needed in the beginning of the impulse response in order to minimize the input-to-output latency. Afterwards, they are increased for lowering the computational complexity. Sometimes, it is beneficial to repeat a sub filter before selecting a longer one, in order to facilitate the reuse of DFT spectra (cp. section 5.2). A consequence of the ascending order is, that sub filters of equal lengths $L_i = L_j$ are neighboring. Such sequences of equally sized sub filters are grouped into *filter segments*. In other words, a filter segment is nothing else than a uniform sub partition. Moreover, any non-uniform filter partition is assembled from such uniform sub partitions.

Frequency representation

All practical non-uniformly partition convolution algorithms realize these segments using uniformly-partitioned convolution techniques (UPOLA or UPOLS). In this respect, it is convenient to regard non-uniform filter partition from another perspective: The focus lies on the filter *segments* and their parameters, instead of the individual *parts* of the partition. It is convenient to write down a non-uniform filter partition using the *frequency representation* of partitions in number theory [7]. The term *frequency* has to be understood by its number theoretic definition, that is the number of repetitions of a specific part. Instead of enumerating the individual parts of a partition (as in Eq. 3.10), all segments are enumerated by their sub filter lengths L_i together with their frequencies P_i (also called *multiplicity*). A partition consisting of M segments is than written down in the following format

$$\mathcal{P} = (\ L_0^{P_0}, \ldots, L_{M-1}^{P_{M-1}}\ ) \tag{6.1}$$

The sub filter lengths $L_i \in \mathcal{S}$ are selected from a set of feasible sub filter sizes $\mathcal{S}$. They form the non-decreasing sequence

$$\forall\, 0 \leq i < M: \quad L_i \leq L_{i+1} \tag{6.2}$$

The frequencies $P_i \in \mathbb{N}_0$ can in principle be arbitrarily chosen. For that the resulting convolution algorithm is real-time capable, they must obey certain causality conditions, which are described in section 6.4.3. Figure 6.2 visualizes the introduced filter partitioning scheme. The notions defined in section 3.2 are now reformulated for the frequency representation in Eq. 6.1.

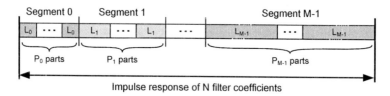

Figure 6.2.: Non-uniform filter partitioning scheme. A filter impulse response is decomposed into a sequence of uniform sub partitions, referred to as filter segments.

The offset of the ith filter segments is the accumulated length up to the ith segment, given by

$$\mathbf{offset}(i) := \sum_{j<i} L_j P_j \qquad (6.3)$$

The total covered filter length of the partition is

$$N = \mathbf{offset}(M) = \sum_{i=0}^{M-1} L_i P_i \qquad (6.4)$$

6.4. Basic algorithm

All non-uniformly partitioned convolution algorithms have a common root: They originate from the filter partitioning introduced in section 3.2 in chapter 3. On the top-level they have a common block diagram, which is depicted in figure 6.3. As before, the audio stream is processed in blocks of B samples. Each length-B input block is stored in a time-domain buffer. This buffer has two purposes: Firstly, it is used to accumulate input samples and repack them into blocks of larger granularity, serving the sub convolutions of the different filter segments. Secondly, by storing previous input data it also allows postponing the computation of these sub convolutions. Its size (capacity) is chosen accordingly to these purposes. Typically, it is implemented as a ring buffer. Each filter segment is realized using an independent uniformly partitioned convolution. The ith filter segment consists of P_i length-L_i sub filters, which are implemented using individual block lengths B_i. The results of all sub convolutions are finally overlap-added in the time-domain. The mixing buffer is dimensioned similar to the input buffer and also implemented as a ring buffer. The sub convolutions are subject to timing dependencies. Usually, a scheduler plans and manages their execution and ensures that all required partial outputs are ready within the deadlines. Some partitioning schemes can be realized using a static execution plan. More details of these aspects are discussed in section 6.8. Before a filter impulse response

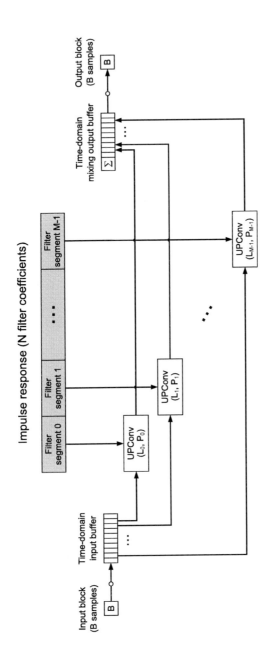

Figure 6.3.: Block diagram of a non-uniformly partitioned convolution algorithm.

can be used with a non-uniform convolution method, it must first be transformed into a frequency-domain representation. Therefore, it is decomposed according to the filter partitioning scheme. Then, the sub filter impulse responses are transformed by the uniformly partitioned convolution algorithms used to implement the segments. Further details on a time-varying filtering are given in section 6.11. Different termini are found in the literature on non-uniform filtering techniques. It is referred to as *non-uniformly partitioned Overlap-Add (NUPOLA)* or *non-uniformly partitioned Overlap-Save (NUPOLS)* convolution, depending on the convolution scheme used for the segments. These notions are as well used in [76]. García [37] uses the term '*multiple-FDL convolution*' for algorithms with freely adjustable partitions.

During the filtering, each sub filter (filter segment) processes the *entire* input stream $x(n)$, but in different granularities (block lengths B_i). Depending on its block length B_i, a segment computes more or less frequently in time, with a rate of computation given by

$$f_i = f_s/B_i \quad [\text{Hz}] \tag{6.5}$$

Segments implemented with small block lengths compute more frequently. Their computational complexity is comparably little. In contrast, later segments realized using large block lengths compute less often. Their computation however can be extensive and even exceed the limited time budget B/f_S of the audio stream. Such computations must therefore either be implemented autonomously, in independent threads, or decomposed into several procedural steps of smaller computational complexity. The very first segment is always processed directly in the audio stream callback routine. To result at minimal input-to-output latency, this leading segment of the partition must use the same block length as the audio stream (see section A.1 in the appendix)

$$B_0 = B \tag{6.6}$$

6.4.1. Standard parameters

Almost all known non-uniformly partitioned convolution algorithms [31, 39, 71, 37, 76, 9] implement the sub filters (segments) using conventional UPOLA/UPOLS convolutions, as introduced in section 5.2. Their block lengths, sub filter lengths and transform sizes are depending as follows:

(1) $\quad B_i = L_i \tag{6.7}$

(2) $\quad K_i = 2B_i = 2L_i \tag{6.8}$

Typically, all sub filter block lengths B_i, including the audio stream block length B, are restricted to powers-of-two [31, 39, 37]

$$(3) \quad B, B_i \in \{\, 2^i \mid i \in \mathbb{N} \,\} \quad \overset{(6.7),(6.8)}{\Longrightarrow} \quad L_i, K_i \in \{\, 2^i \mid i \in \mathbb{N} \,\} \qquad (6.9)$$

This is not only reasonable for the fast computation of the transforms. The order in Eq. 6.2 implies, that all entities B_i and L_i are multiples of the audio stream block length B

$$(4) \quad B_i = 2^p, B_{i+1} = 2^q \;\; (p, q \in \mathbb{N}) \quad \overset{(6.2)}{\Longrightarrow} \quad p \le q \quad \overset{(6.6)}{\Longrightarrow} \quad B \mid B_i \qquad (6.10)$$

$$(5) \quad B \mid B_i \quad \overset{(6.7)}{\Longrightarrow} \quad B \mid L_i \quad \overset{(6.3)}{\Longrightarrow} \quad B \mid \mathbf{offset}(i) \qquad (6.11)$$

In particular the relations in Eq. 6.11 greatly simplify the scheduling of sub convolutions, as the next section will show.

6.4.2. Computational complexity

The computational costs of a non-uniformly partitioned convolution algorithm are derived from the block diagram in figure 6.1. As each parallel sub convolution processes the complete audio stream, the overall computational costs are made up from the costs of all sub filters. Additionally, the sub filter results must be accumulated into the overall output signal. This requires an extra addition per output sample for each sub filter. The computational overhead of the scheduling of the sub convolutions is usually left out of the consideration. In the following a standard implementation with conventional UPOLS convolutions is regarded.

Let $\mathcal{P} = (\, L_0^{P_0}, \ldots, L_{M-1}^{P_{M-1}} \,)$ be a non-uniform filter partition. The computational costs (per output sample) of the stream processing are given by

$$T_{\text{NUPOLS}}^{\text{stream}}(\mathcal{P}) := \sum_{i=0}^{M-1} \left(T_{\text{UPOLS}}^{\text{stream}}(B_i, L_i \cdot P_i) + \frac{T_{\text{ADD}}(B_i)}{B_i} \right) \qquad (6.12)$$

The term $T_{\text{UPOLS}}^{\text{stream}}(B_i, L_i P_i)$ corresponds to the costs per sample for an UPOLS convolution with equal block length and sub filter length $B_i = L_i$, consisting of P_i parts. $T_{\text{ADD}}(B_i)/B_i$ expresses the additional costs per sample for overlap-adding the segments' results into the global output. This portion is often neglected. A filter transformation of the complete length-N impulse response $h(n)$ requires to transform all sub filter impulse responses $h_i(n)$

$$T_{\text{NUPOLS}}^{\text{ftrans}}(\mathcal{P}) := \sum_{i=0}^{M-1} T_{\text{UPOLS}}^{\text{ftrans}}(B_i, L_i \cdot P_i) \qquad (6.13)$$

A very important property of both cost functions is, that they are *separable*. Let T be synonymous for the two cost functions (Eq. 6.12 or 6.13) and let the expression $T(L_i^{P_i})$ correspond to the elements in their right-hand side sums. Then both functions fulfill

$$T(\mathcal{P}) = T(L_0^{P_0}) + \cdots + T(L_{M-1}^{P_{M-1}}) \tag{6.14}$$

Analyzing both cost functions, two key factors of the computational costs can be identified: Firstly, the underlying uniformly partitioned convolution of choice (responsible for the sub filter costs T_{UPOLS}). Any improvements here, lower the computational costs of the non-uniform method as well. Secondly, the filter partition $\mathcal{P}$. The choice of a suitable non-uniform partition is not trivial and subject of the subsequent considerations. Before the computational costs of non-uniformly partition convolution techniques are evaluated, different filter partitioning techniques are reviewed first.

6.4.3. Timing dependencies

The non-uniform filter partition introduces timing dependencies for the sub filter convolutions. Points in time exist, when a sub filter *can* begin its computation (availability of input samples) and when its computation *has* to be finished (deadline for output samples), in order to assemble the overall convolution output. The computation of the corresponding sub filter convolutions has to be finished within these limited time intervals, otherwise partial results are missing in the overall output. Given the case that the admissible time interval exceeds the actual computation time, the processing of the sub filter can be shifted along the time axis. The amount of flexibility thereby is considered as the *clearance* and formalized in the next paragraphs. The timing-dependencies with result for a partition can be visualized by diagrams, as exemplary shown in figure 6.4.

Some partitions can even inhere in insufficient time spans and are thus not realizable in real-time. A non-uniform partition is only real-time compliant if it fulfills fundamental conditions, which guarantee, that all involved sub filter computations can be finished in time. Egelmeers and Sommen [30] derived such conditions first, for a sample-wise processing. The time scale in their work is quasi-continuous and can be measured in seconds, clock cycles or samples $t \in \mathbb{R}^+$.

This work focuses on block-based processing, where the events in audio processing align with a quantized times $t_n = nT + T_0$ (see section A.1). The offset T_0 of the scale is irrelevant for the subsequent timing dependencies and can be dropped. Consequently, it is sufficient and convenient to describe times just by stream cycles n instead of a times t_n. Actual computation times

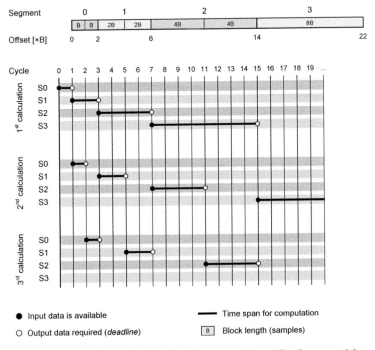

Figure 6.4.: Timing dependencies for the non-uniform Gardner partition
$$\mathcal{P} = (\,[B]^2, [2B]^2, [4B]^2, [8B]^1\,)$$

are not considered. It is assumed that the clearance is sufficiently large, so that each sub convolution can be computed in the given time span.

The derivation of timing dependencies was published by the author in [121]. The considerations here are developed in a more general way, also regarding individual iterations of the sub convolutions. First, these general dependencies are derived for arbitrary filter partitions. Afterwards, the results are limited to the standard parameters, expressed in Eq. 6.7-6.11, as found in [121].

General dependencies

Let $\mathcal{P} = (\,L_0^{P_0}, \ldots, L_{M-1}^{P_{M-1}}\,)$ be a filter partition consisting of $M \in \mathbb{N}_0$ segments. In the following the filter segment of index $0 \leq i < M$ is regarded. It is assumed that this segment $L_i^{P_i}$ is implemented using an arbitrary block length B_i. Furthermore, let the index $m \in \mathbb{N}_0$ denote the mth iteration (execution) of the sub convolution. In order to make things easier understandable, this index begins with 1 ($m \leftrightarrow m^{\text{th}}$ execution).

The segment can start its computation the mth time, as soon as mB_i input samples have been accumulated. This holds for cycles with an index k which fulfills the following inequality

$$mB_i \leq (k+1)B \qquad (m \in \mathbb{N}_0, k \in \mathbb{N}) \tag{6.15}$$

According to the semantics defined in section A.1 in the appendix, $(k+1)B$ input samples have been provided in the cycle k of the audio stream. Solving Eq. 6.15 for the smallest possible value of k, determines the first cycle for which the mth computation can start. This defines the point in time

$$\mathbf{avail}^{(m)}(i) := \left\lceil m \frac{B_i}{B} \right\rceil - 1 \tag{6.16}$$

Trivially, the first segment implemented with the stream block length $B_0 = B$ can compute every cycle. Given that $B \mid B_i$, the ceiling function can be omitted. In this case the computation can start exactly every m^{th} cycle. In other words, if $B \nmid B_i$ the result is an irregular availability, which can lead to an uneven load distribution.

The *deadline* for a sub convolution is defined as the point in time, when the output of a sub filter contributes to the overall result. Here, the computation of the sub filter must strictly be finished. This includes the overlap-and-add step in the final output. This definition of the deadline has the advantage of being independent of the computation time, which it does not account for.

The point in time (measured in samples) when the segment $L_i^{P_i}$ contributes to the overall convolution the *first time*, is determined by its position $\mathbf{offset}(i)$ (Eq. 6.3) in the impulse response. Recall, that all sub filters process the entire audio stream, but with different granularities (block lengths B_i). Hence, the mth computation of the segment i contributes to the output $(m-1)B_i$ samples later. For a sample-wise processing (cp. [30]), the offset defines the deadline

$$\mathbf{offset}(i) + (m-1)B_i \tag{6.17}$$

The considered block-based processing aligns with the stream cycles, not individual samples. Applying Eq. A.2 translates Eq. 6.17 from the index of samples into the index of cycles. This defines the deadline for the mth computation of the segment as

$$\mathbf{deadline}^{(m)}(i) = \left\lfloor \frac{\mathbf{offset}(i) + (m-1)B_i}{B} \right\rfloor \tag{6.18}$$

Given that the block length B divides the numerator in Eq. 6.18, the floor function can be dropped. In this case the deadlines form a uniformly-spaced sequence in time. Otherwise the pattern of deadlines is irregular.

The admissible time span for the mth computation of the ith segment is the span from the first point in time, when the computation *can* be performed to the point in time, when it *must* have been performed. It is defined as the difference in cycles

$$\text{clearance}^{(m)}(i) \;=\; \text{deadline}^{(m)}(i) - \text{avail}^{(m)}(i) \qquad (6.19)$$

The term '*clearance*' was favored by the author for a specific reason: It indicates the *tolerance* in time for the computation of a filter segment. A *zero-clearance* ($\text{clearance}^{(m)}(i) = 0$) marks a tight fit in time. Here, the input data is available just within the same cycle that the output samples must be provided ($\text{deadline}^{(m)}(i) = \text{avail}^{(m)}(i)$). The computation time of the segment is therefore pinned to a *single* cycle and cannot be shifted to a later point in time. Zero-clearances mark critical timing dependencies. Positive clearances ($\text{deadline}(i) > \text{avail}(i)$) allow for a deferred computation and provide more flexibility to the scheduling of sub convolutions.

The maximal number of cycles, that the mth computation of the segment can stretch over is

$$\text{timespan}^{(m)}(i) \;=\; \text{deadline}^{(m)}(i) - \text{avail}^{(m)}(i) + 1 \qquad (6.20)$$
$$=\; \text{clearance}^{(m)}(i) + 1 \qquad (6.21)$$

Standard parameters

Running the sub filters with standard parameters, as defined in Eq. 6.7-6.11 simplifies the above stated timing measure significantly. All events for the computation of sub filters align with the processing of the stream. The ceiling and floor functions can be dropped, as the numerators in Eq. 6.16 and Eq. 6.18 are dividable by B. Both, the possible start times and deadlines, repeat every B_i/B cycles and are hence uniformly-spaced over time. Consequently, the clearance of each segment becomes independent of the actual iteration m. This result in particular simplifies the scheduling of sub convolutions enormously. For the sake of simplicity, the iteration index m is omitted from the other two terms as well. The following simplified timing dependencies can be summarized for choice of standard parameters defined in Eq. 6.7-6.11

$$\mathbf{avail}(i) = B^{-1}B_i - 1 = B^{-1}L_i - 1 \tag{6.22}$$

$$\mathbf{deadline}(i) = B^{-1}\mathbf{offset}(i) \tag{6.23}$$

$$\mathbf{clearance}(i) = \mathbf{deadline}(i) - \mathbf{avail}(i)$$

$$= B^{-1}\left[\left(\sum_{j<i} L_j P_j\right) - L_i + B\right] \tag{6.24}$$

The unit of all three measures are cycles, starting from 0. They can be easily converted back into the scale of samples by multiplying them with B. Unless outlined, all subsequent considerations are based on the definitions in Eq. 6.22-6.24.

6.4.4. Causal partitions

Negative values $\mathbf{clearance}(i) < 0$ signalize, that the computation of the segment has to be finished before the input data becomes available ($\mathbf{deadline}(i) < \mathbf{avail}(i)$). Filter partitions containing segments with this property can thus not be realized under real-time conditions. They are referred to as *non-causal* partitions and are not of practical interest.

Definition: A filter partition $\mathcal{P} = (\ L_0^{P_0}, \ldots, L_{M-1}^{P_{M-1}}\ )$ is *causal*, if and only if all of its segments satisfy $\forall 0 \leq i < M : \mathbf{clearance}(i) \geq 0$.

Combining this condition with the inner term in the right-hand side of Eq. 6.24 yields

$$\mathbf{clearance}(i) \geq 0 \quad \overset{(6.24)}{\Longleftrightarrow} \quad \left(\sum_{j<i} L_j P_j\right) - L_i + B \geq 0 \tag{6.25}$$

$$\Longleftrightarrow \quad L_i \leq \left(\sum_{j<i} L_j P_j\right) + B \tag{6.26}$$

Eq. 6.26 provides a fundamental rule for the construction of causal filter partitions. Let $\mathcal{P} = (\ L_0^{P_0}, \ldots, L_{M-1}^{P_{M-1}}\ )$ be a causal filter partition consisting of M segments. Then $\mathcal{P}$ can be extended to a partition $\mathcal{P}' = (\ L_0^{P_0}, \ldots, L_M^{P_M}\ )$ with $M+1$ segments, by appending a further segment $L_M^{P_M}$. Condition 6.26 states, that $\mathcal{P}'$ is causal if and only if the appended segment has a sufficiently small sub filter length L_i. In other words, condition 6.26 limits the growth of the sub filter lengths L_i over the filter length N. It is further worth pointing out, that if a partition $\mathcal{P}$ is causal, the causality is preserved by increasing its multiplicities P_i arbitrarily. The contrary however does not hold. Reducing the multiplicity P_i of a segment in $\mathcal{P}$ may violate the causality.

6.4.5. Canonical partition

The fastest possible increase in sub filter lengths is achieved by claiming equality in condition 6.26. This instantaneously implies that all clearances will be zero. Then a unique partition for every filter length N can be constructed as follows: Let $\mathcal{S}$ be a set of viable sub filter sizes $L_i \in \mathcal{S}$ and furthermore let $B \in \mathcal{S}$. The first segment is chosen $L_0 = B, P_0 = 1$ as usual. Then parts are appended successively, until the length N is reached. Thereby, each sub filter length $L_i \in \mathcal{S}$ is maximized according to condition 6.26. Given, that no $L_i \in \mathcal{S}$ fulfills condition 6.26, the multiplicity P_{i-1} of the preceeding segment is increased until such an $L_i \in \mathcal{S}$ is found. With respect to the selection of $\mathcal{S}$ the construction is unique for a filter length N. Such a non-uniform filter partition is referred to as the *canonical partition* of length N and $\mathcal{S}$.

Considering sub filter lengths $L_i \in \mathcal{S} = \{B, 2 \cdot B, 2^2 \cdot B, 2^3 \cdot B, \dots\}$ (power-of-two multiples of the audio stream block length B), a unique construction of the following form is found

$$L_0 = B$$
$$L_1 = L_0 + B = 2B$$
$$L_2 = L_0 + L_1 + B = 4B$$
$$L_3 = L_0 + L_1 + L_2 + B = 8B$$

$$\vdots$$

$$L_M = \left(\sum_{i=0}^{M-1} 2^i B \right) + B = 2^M B \tag{6.27}$$

This defines a non-uniform filter partition of the form

$$\mathcal{P}_{\text{canon}} = \left([B]^1, [2B]^1, \dots, [2^k B]^1 \right) \quad (k \in \mathbb{N}) \tag{6.28}$$

Each segment consists of a single part only ($P_i = 1$). Hence, the overall number of individual filter *parts* is minimal. This can easily be shown by proof with contradiction. However, the number of filter *segments* is not necessarily minimal. A counter example is the uniform partition, which only consists of a single segment.

6. Non-uniformly partitioned convolution

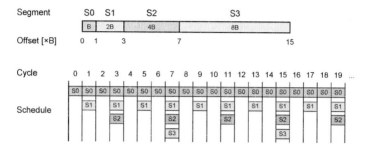

Figure 6.5.: Schedule of the sub convolutions in a canonical partition.

A canonical partition, as defined in Eq. 6.28, has the following attributes

$$\mathbf{avail}(i) = B^{-1}(2^i B) - 1 = 2^i - 1 \tag{6.29}$$

$$\mathbf{offset}(i) = \sum_{j=0}^{i-1} 2^j B = B(2^i - 1) \tag{6.30}$$

$$\mathbf{deadline}(i) = B^{-1}\mathbf{offset}(i) = 2^i - 1 \tag{6.31}$$

$$\mathbf{clearance}(i) = \mathbf{deadline}(i) - \mathbf{avail}(i) = 0 \tag{6.32}$$

$$\mathbf{timespan}(i) = \mathbf{clearance}(i) + 1 = 1 \tag{6.33}$$

The canonical partition embodies the principle of the most rapid increase in sub filter sizes. One could presume, that this strategy results in the lowest computational costs. It is shown later, that this is not the case and that partitions with a lower computational complexity exist. As each segment consists of only a single part, the corresponding sub convolution is essentially unpartitioned and does not facilitate the reuse of DFT spectra. Worse though, the canonical partition implies a worst-case load distribution. Figure 6.6(a) shows the occurring computational load (required CPU cycles) for the individual cycles of the audio processing. Computations can not be distributed along the time axis (zero clearances) and all sub filter convolution can superpose in certain cycles. The results in distinct load peaks, which severely limit the maximal performance.

6.4.6. Gardner's partitioning scheme

The above stated drawbacks of the canonical partition make it completely impractical. Gardner [39] pursued the concept of the most rapid increase in sub filter sizes and brought it to the level of practicality, by introducing a couple of modifications. His partitioning scheme is outlined in the following and its properties are discussed. However, the later results by García [37] showed that the principle does not guarantee a *minimal-cost decomposition*, as mistakenly supposed by [39]. In his work, Gardner [39] regards a mixed strategy to achieve an even lower latency: The leading sub filter is implemented using a sample-wise time-domain FIR filter (e.g. in DSP hardware). As the focus lies on the block-processing here, these aspects are left out of the considerations.

In order to overcome the inevitable implications of zero-clearances, each segment is implemented with a multiplicity of two, instead of one. Only the last segment may have a multiplicity of one or two. This results in non-uniform filter partitions of the following form

$$\mathcal{P}_{\text{Gardner}} = \left([B]^2, [2B]^2, \ldots, [2^{k-1}B]^2, [2^k B]^l \right) \quad (k \in \mathbb{N}, l \in \{1, 2\}) \quad (6.34)$$

In this work such partitions are referred to as *Gardner partitions*. The (block-based) sub filters are staggered in powers-of-two, with parameters $L_i = 2^i B, P_i = 2, P_{M-1} \in \{1, 2\}$. The segments of a Gardner partition have the following properties

$$\textbf{avail}(i) = B^{-1}(2^i B) - 1 = 2^i - 1 \quad (6.35)$$

$$\textbf{offset}(i) = \sum_{j=0}^{i-1} 2 \cdot 2^j B = B \sum_{j=0}^{i-1} 2^{j+1} = B(2^{i+1} - 2) \quad (6.36)$$

$$\textbf{deadline}(i) = B^{-1}\textbf{offset}(i) = 2^{i+1} - 2 \quad (6.37)$$

$$\textbf{clearance}(i) = \textbf{deadline}(i) - \textbf{avail}(i) = 2^{i+1} - 2 - 2^i + 1 = 2^i - 1 \quad (6.38)$$

$$\textbf{timespan}(i) = \textbf{clearance}(i) + 1 = 2^i \quad (6.39)$$

The design principle of this partitioning scheme is, that each sub convolution is granted a admissible time span for its computation, that equals the block length of the sub convolution. A clearance of zero indicates that the computation can maximally span over a single cycle. Accordingly, a clearance of k cycles corresponds to a maximal admissible computation time of $(k+1)B/f_S$. Assuming, that the computation of each segment can be evenly

163

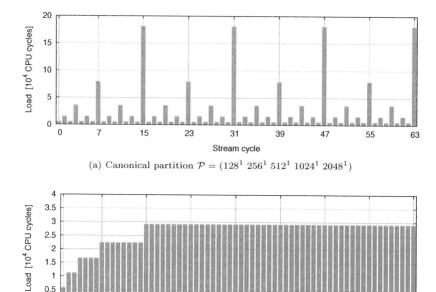

(a) Canonical partition $\mathcal{P} = (128^1\ 256^1\ 512^1\ 1024^1\ 2048^1)$

(b) Gardner partition $\mathcal{P} = (128^2\ 256^2\ 512^2\ 1024^2\ 2048^1)$

Figure 6.6.: Computational load per stream cycle for two partitions on the test system

distributed over this admissible time span, the load distribution of the Gardner algorithm becomes uniform. This is visualized in figure 6.6(b). The costs increase until cycle 15, from which on all segments continuously compute in parallel. Comparing this result to the uneven load distribution of a canonical partition in figure 6.6(a), the importance of scheduling flexibility for the sub convolutions becomes obvious. However, achieving an even load distribution as in figure 6.6(b) on an actual system is difficult. Splitting up the convolution of a larger sub filter into several portions of similar computational effort is challenging, e.g. due to the mixture of different operations (FFTs and spectral convolutions). These implementation-specific issues are addressed in section 6.8.

6.5. Optimized filter partitions

The preceding sections introduced several non-uniform filter partitions and examined their properties. Schemes, like Gardner's partitioning [39], follow a certain design principle and define a formal way of decomposing a filter impulse response of arbitrary length N. However, numerous different possibilities of partitioning a filter exist. All of these have different computational costs and properties. Given a cost function $T(\mathcal{P})$, the partition can be identified that minimizes the computational costs. However, this does usually not result in a satisfactory solution. Optimality is always subject to the postulated optimization goals. Particularly for non-uniform filter partitions, multiple aspects must be carefully considered. A partition which results in the lowest computational costs is worthless if it cannot be realized in real-time (e.g. a non-causal partition). Even if it can be realized and the computational costs are minimal, the load distribution can be so unfavorable, that a partition with larger computational effort is actually more practical.

This section reviews approaches to obtain non-uniform filter partitions as solutions of an optimization problem. First, the formal optimization problem is introduced. Afterwards, the optimization algorithm by García [37] is presented. This algorithm facilitates dynamic programming to solve the problem in polynomial runtime. It is shown, that the resulting minimal-cost partitions inhere no clearances and are practically unfavorable. Finally, enhancements to the optimization algorithm are presented, which allow considering further aspects in the optimization procedure, like timing dependencies and scheduling of sub convolutions.

6.5.1. Optimization problem

In the following, the search for an optimal non-uniform filter partition is regarded as a formal optimization problem. Let $\mathcal{F}$ denote a set of feasible non-uniform partitions. For the standard NUPOLS technique (cp. Sec. 6.4.1), this is usually the set of all causal partitions consisting of sub filters, whose lengths L_i are powers-of-two (Eq. 6.9). Let $T : \mathcal{F} \to \mathbb{R}^+$ be a cost function for the partitions in $\mathcal{F}$. This is typically the costs per output sample $T_{\text{NUPOLS}}^{\text{stream}}(\mathcal{P})$ (Eq. 6.12), that result from a specific partition $\mathcal{P}$. Usually, the interest lies in the minimal possible costs, defining the following minimization problem: The optimal filter partition $\mathcal{P}_{\text{opt}} \in \mathcal{F}$ is defined as the one that minimizes the function $T(\mathcal{P})$

$$\text{cost}_{opt} := \min_{\mathcal{P} \in \mathcal{F}} T(\mathcal{P}) \qquad \mathcal{P}_{opt} := \operatorname*{argmin}_{\mathcal{P} \in \mathcal{F}} T(\mathcal{P}) \qquad (6.40)$$

6.5.2. Optimization algorithm

In 2002, García [37] presented a formalism to represent the construction of filter partitions by states in a discrete state-space. This allows mapping the set of feasible partitions to a weighted directed acyclic graph (DAG). The optimization problem is thereby reformulated as a shortest-path problem, that can be efficiently solved using the Viterbi algorithm [93] in polynomial runtime. The principle is briefly reproduced here. For deeper explanations, the reader is referred to [37].

State-space representation

The state-space representation in this thesis is based on García's original representation [37] with a few modifications. These are necessary, as García's work considers a sample-wise processing, as Gardners approach [39]. This manifests in a slightly different formulation of the causality condition $L_i \leq$ offset(i). Conceptually, both mappings are similar. Additionally in this thesis, states are represented by triples of integers, including also the offset as a variable. Figure 6.7 visualizes the construction process of a non-uniform filter partition as suggested in [37]. One has to imagine a *pointer*, which iterates over the filter length in steps of the block length B. In Fig. 6.7 this pointer (zone) is marked with gray background. With each iteration, the previous partitions are combined into partitions of larger lengths. Vertically separated areas ($\geq B, \geq 2B, \cdots$) correspond to the iterations in the construction. After the nth iteration, partitions of length nB or longer have been constructed.

Each state thereby corresponds to a sub filter *placement* in a partition. States are labeled in the form of triplets ($N.S.Q$). All three variables are interpreted as multiples of the block length B. N marks the right margin (end) of the sub filter ($\times B$). This corresponds to the offset of any succeeding sub filter. S denotes the length of the sub filter ($\times B$). Q marks the fraction of the sub filter (half, fourth, eighth, etc.), that coincides with the current pointer range (marked gray). The state (7.4.3) corresponds to a sub filter of length $4B$, whose third quarter aligns with the pointer region and its right margin is located at $7B$. The initial state of the DAG is (1.1.1) and describes the trivial partition (B^1). Figure 6.8 shows the resulting DAG (without weights) for filter lengths $N \geq 7B$.

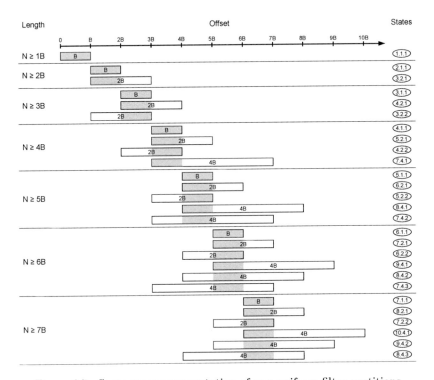

Figure 6.7.: State-space representation of non-uniform filter partitions

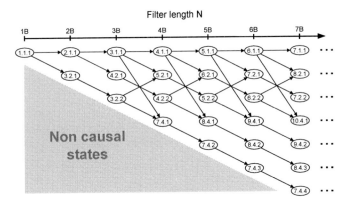

Figure 6.8.: DAG representation of feasible non-uniform filter partitions

167

Transitions

Three types of possible transitions between states can be identified

1. Continuation of a sub filter: $(N.X.Q) \rightarrow (N.X.Q+1)$ with $Q < X$
2. Repetition of preceding sub filter: $(N.X.X) \rightarrow (N+X.X.1)$
3. Transition to a longer sub filter: $(N.X.X) \rightarrow (N+Y.Y.1)$ with $Y > X$

The causality condition in Eq. 6.26 can then be expressed by means of the state variables

$$\text{clearance}(i) = B^{-1}\text{offset}(i) - B^{-1}L_i + 1 \geq 0 \qquad (6.41)$$
$$\Updownarrow \qquad \Updownarrow$$
$$\text{clearance}(i) = \qquad N \quad - \quad S \quad + 1 \geq 0 \qquad (6.42)$$

Consequently, the DAG representing causal partitions, does not contain states (N, N, Q), as these violate condition 6.42. Moreover, the principle of causality (see section 6.3) only affects the transition of type 3. Transitions of type 1 and 2 can always be performed without restrictions.

Weights

Each transition in the DAG is attributed with the costs of the corresponding extension. In the following, the weights are derived from the computational costs per output sample of the non-uniformly partitioned Overlap-Save (NUPOLS) algorithm, defined in Eq. 6.13 and more specifically Eq. 5.6. The evaluation of costs is thereby separated according to the construction process and regarded for each transition.

The initial state $(1, 1, 1)$ represents a segment consisting of a single element only. Its computation involves a forward and inverse FFT, a spectral convolution and final overlap-add step in the NUPOLS algorithm

$$\frac{1}{B}\Big[T_{\text{FFT}}(2B) + T_{\text{CMUL}}(B + 1) + T_{\text{IFFT}}(2B) + T_{\text{ADD}}(B)\Big] \qquad (6.43)$$

Trivially, any ϵ-transition of type 1 does not affect the costs. Repetitions of type 2 correspond to an increase of a segment, which already existed. The additional effort is a further spectral convolution and accumulation

$$\frac{1}{B}T_{\text{CMAC}}(B + 1) \qquad (6.44)$$

Any transition to a larger sub filter $B_i = Y \cdot B$ (type 3) opens a new segment of this length, resulting in costs

$$\frac{1}{B_i}\Big[T_{\mathrm{FFT}}(2B_i) + T_{\mathrm{CMUL}}(B_i + 1) + T_{\mathrm{IFFT}}(2B_i) + T_{\mathrm{ADD}}(B_i)\Big] \qquad (6.45)$$

As each partition corresponds to a path in the DAG, its costs are accumulated from the weights of all edges on this path.

Viterbi algorithm

The optimal partition problem can be solved using dynamic programming (DP), because the cost function is separable (see Sec. 6.4.2) and thus fulfills Bellmann's *principle of optimality* [25]. The optimal partition can thus be found using the *Viterbi algorithm* [93] in polynomial runtime (PTIME). Details on the implementation specifically for the optimal partition problem are given in [37]. The algorithm needs to store the states of only two iterations (current and next), keeping the memory footprint manageable. Memorizing all the states is impossible, due to the exponentially growing memory requirements. First, the Viterbi algorithm tracks down the optimal sequence to a final state of the desired target length N. This also determines only the minimal costs, but not the complete partition. Afterwards, *back tracking* [93] is applied to determine also the optimal partition itself, which is provided as a sequence of states. The author created and benchmarked a C++ implementation of the optimization algorithm. The execution times range from seconds to minutes on a standard PC for all relevant problem sizes (e.g. filters up to 10 s of reverberation time). The results of the optimization procedure are discussed in the following.

6.5.3. Minimal-load partitions

In essence, minimal-load partitions facilitate the principle of the most rapid increase of sub filter lengths, but by fully utilizing the potential of the FDL. The increased multiplicities prove, that the latter is actually exploited. In contrast to canonical partitions and Gardner's partitioning scheme, not all possible power-of-two sub filter lengths are involved. Instead, minimal-load partitions select even larger sub filter lengths for a next segment, as soon as the savings on an FDL are exhausted. Still, this solution results in zero clearances for most of the segments. This result was published by the author in [121].

The complete list of minimal-load partitions for a block length $B \in \{128, 256, 512\}$ and filter length $B \leq N \leq 512k$ are printed in sections A.2.1, A.2.2 and A.2.3 in the appendix. From these results, the following properties of minimal-load partitions can be summarized.

- Sub filters of length L_i up to 64k are involved, but not longer.

- Sub filters multiplicities range up to 32, proving that the FDL concept is extensively utilized.

- At maximum four filter segments are used, but not more.

- A uniform partitioning marks the optimal solution if the filters are not too long. This roughly holds within the range $20B \leq N \leq 30B$.

- For typical filter length in artificial reverberation, a three segment layout is optimal. If the block lengths get longer, a two segment partition is sufficient.

- The majority of filter segments have zero-clearances. This always holds for leading segments. Exceptions are only found in later segments. Shorter block lengths (latencies) emphasize the problem.

Particularly the last observation is very important. It marks a common property of minimal-load partitions, not only in practical examinations (benchmarks), but also for theoretical runtime complexities (see [121]). The zero-clearances in minimal-load partitions render them unusable in practice, due to a similarly unfavorable load distribution as for the canonical partition (cp. Fig. 6.6(a)). In other words, a minimal-load partition for a filter length N requires the least number of operations, but its computation load is badly distributed. In fact, the resulting load peaks form the bottleneck for the computation. An optimization only aiming at the least computational effort does not produce usable results. This proves evidence, that the optimization must carefully consider further aspects. The following section describes strategies to accomplish this. The analysis of the computational load is found in section 6.6.

6.5.4. Practical partitions

The author proposed a solution to the problem in [121], which is shortly outlined in the following. For the detailed study of the results the reader is referred to [121]. The basic idea is to enforce certain minimal clearances for all possible segments. This can be seen as a restriction of the set of feasible solutions $\mathcal{F}$. Then the optimizer decides on the optimal partition within this restricted search space.

Search space restrictions

Instead of allowing all causal (i.e. realizable) partitions, further restrictions are introduced. They are methodically similar to the causality condition (Eq.

6.26) and limit the increase in sub filter lengths during the construction of partitions. The result is a specification of required minimal clearances. For each potential filter segment $L_i^{P_i}$ a minimal clearance can be postulated

$$\mathbf{clearance}(i) \geq \mathbf{clearance}_{\min}(i) \qquad (6.46)$$

These conditions are transformed into restrictions on states in the DAG construction, similar to Eq. 6.41,6.42. Not only the minimal clearance can be specified, but also minimal and maximal multiplicities for the segments (details see [121]).

Restrictions

Introducing restrictions on the search space, usually increases the computational costs. The minimal clearance conditions prevent the optimization to select larger sub filters early in the construction (at the beginning). In a minimal-load partition the sub filter length is increased, if this minimizes the computational costs. On the contrary, minimal clearance conditions can shift such decisions to later points in time. As there are only two design choices—*repetition* of a sub filter or selection of a *longer* one—the minimal-clearances thus increase the multiplicities over those in minimal-load partitions. According to the considerations in section 6.4.3, these do break the causality, but increase the clearances and thus the scheduling flexibility. A general relationship is observed in section 6.6: The more scheduling flexibility a non-uniform filter partition inheres, the higher are its costs compared to the minimal-load partition. Which restrictions are necessary depends on the actual implementation. Section 6.8 addresses these issues. Section 6.12 draws some conclusions on meaningful restrictions for the application of room acoustic auralization.

The author proposed two example restriction sets with practical relevance in [121], which are explained in the following:

1. Each filter segment $L_i^{P_i}$ is granted at least $B \cdot \mathrm{timespan}(i) \geq L_i$ (cp. Eq. 6.21) for its computation. Expressed in clearances this reads: $\mathrm{clearance}_{\min}(i) \geq L_i/B - 1$. This strategy follows the same principle as Gardner's approach [39] does (see Sec. 6.4.6), but explicitly allows FDLs to be exploited.

2. This strategy is inspired by observations in practice. Segments of sub filter lengths $2B$ and $4B$ are enforced a computation time span of at least two cycles. Segments of sub filter lengths $8B \ldots 64B$ are granted four cycles. All other segments are ensured a time span of eight cycles. These rules are less restrictive than strategy 1. The first segment is only allowed a multiplicity in the range $2 \leq P_0 \leq 4$. This condition

has the following function: First segments have to be processed within the time-limited audio callback. In order to enable a large number of channels to be processed, the individual costs per channel shall not be too large. Otherwise the maximal possible number of channels is limited. Hence, it is beneficial to keep the multiplicity small. On the other hand, the multiplicity shall not be too low either, because in this case there is no clearance for the succeeding segment. These aspects become very important in parallelized implementations where all other segments are computed in background threads (cp. Sec. 6.8).

The optimal partitions for both variants are summarized in sections A.2.4 and A.2.5 in the appendix. The impact of the restriction on the computational effort are examined in section 6.6.

6.6. Performance

The computational costs of the standard NUPOLS are examined for the introduced non-uniform filter partitions. The costs for a partition $\mathcal{P}$ are found by evaluating Eq. 6.12 respectively Eq. 6.13 with the benchmarks of the test system (see Ch. 7). Costs for the stream processing are measured as before, by the number of CPU cycles per filtered output sample. The effort for filter transformations is accounted by the total number of CPU cycles.

Non-uniform vs. uniform partitioning

First, general comparisons between the non-uniformly and uniformly partitioned methods are conducted. These comparisons consider the optimal minimal-load partitions (see Sec. 6.5.3) for the NUPOLS technique. Figure 6.9 shows the resulting computational effort for both approaches (non-uniform=solid lines, uniform=dashed lines) over the filter length N. Like before, several different block lengths B are regarded. The following observerations can be made: Non-uniform techniques have a different and logarithm-like progression of the costs. Doubling the filter length does not double the computational costs any more. The block length influences the costs much less and not with the previously observed anti-proportional behavior. It exists a certain filter length, from which a non-uniform partition is actually used. For smaller filter lengths, both methods converge into a uniform partitioning. These distinct filter lengths can be identified in the list of optimal minimal-load partitions in the appendix (see Sec. A.2.1, A.2.2 and A.2.3). A uniform-partitioning turned out optimal when the filter length does not exceed $20 - 30\times$ the audio stream block length.

Figure 6.12 shows the relative speedup, between the UPOLS and NUPOLS technique. This speedup grows over the filter length N and depends as well on the block length B. In particular for low-latency filtering, the savings by non-uniform partitioning are notably. Considering a room impulse response of 100,000 taps and a stream block length $B = 128$, the NUPOLS filtering outperforms the UPOLS approach by a factor of approximately 10. For a longer block length $B = 1024$ the savings drop a factor of about 2. It can be seen, that non-uniformly partitioned techniques are asymptotically faster than uniformly-partitioned approaches.

Figure 6.12 examines the dependency of the block length and computational costs of the NUPOLS methods. It depicts the relative speedup, when switching from the reference block length $B = 128$ to a larger one. The results strongly differ from the observed behavior for the UPOLS technique in figure 5.6. The costs of a non-uniformly partitioned convolution are much less influenced by the block length. Only for filters of lengths of approximately $N < 50,000$ taps, the change to a larger block causes a significant reduction of the costs (up to a factor of 2.2). For very long filters, the advances by using an eight-fold block length manifest in a speedup ≈ 1.3. Whereas in contrast, a uniform-partitioning would deliver about an eight-fold speedup.

The computational effort of the filter transformation is shown in figure 6.11. Here, a different observation is made: Transforming a non-uniformly partitioned filter is more expensive than transforming a uniformly partitioned one. For very long filters (e.g. 400,000 taps), the transformation for the NUPOLS method can be about 30-40% more expensive than for the UPOLS technique. It is worth mentioning, that the additional expenses are little for short filter lengths. When different parts of the impulse response are exchanged asynchronously (see Sec. 6.12), the additional expenses do not impact the performance to the same extent. Like before, the filter transformation for both approaches is only marginally affected by the block length.

Finally, figure 6.15 visualizes how the computational costs in the NUPOLS are constituted. Compared to the UPOLS technique (see figure 5.7), fast transforms play a much more important role. In the NUPOLS algorithm, FFTs make up the majority of computations, whereas in the UPOLS technique they mark fixed costs, independent of the filter length. From this observation it can be concluded, that the performance of non-uniformly partitioned convolution is strongly affected by the performance of the fast Fourier transforms.

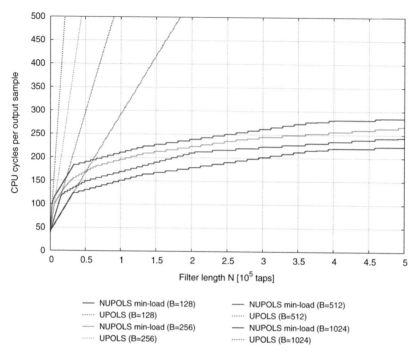

Figure 6.9.: NUPOLS versus UPOLS convolution on the test system: Computational costs of stream processing

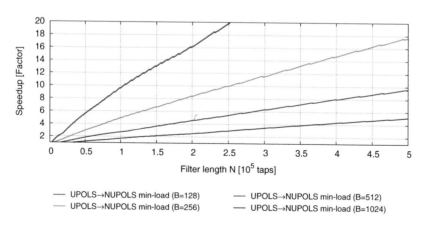

Figure 6.10.: NUPOLS versus UPOLS convolution on the test system: Stream processing speedup over the block length

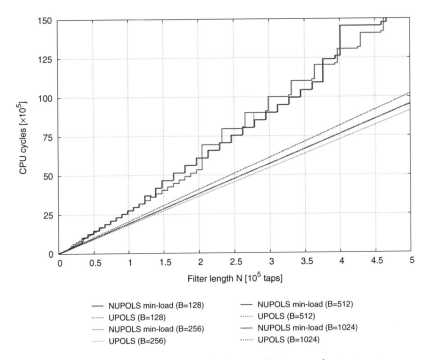

Figure 6.11.: NUPOLS versus UPOLS convolution on the test system:
Computational costs of a filter transformation

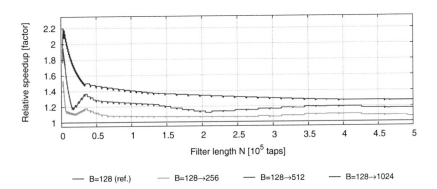

Figure 6.12.: NUPOLS convolution on the test system: Relative computational costs of the stream processing for different block lengths

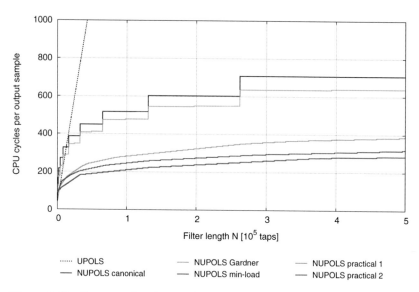

Figure 6.13.: Computational costs for the stream processing of different convolution techniques on the test system (block length $B = 128$).

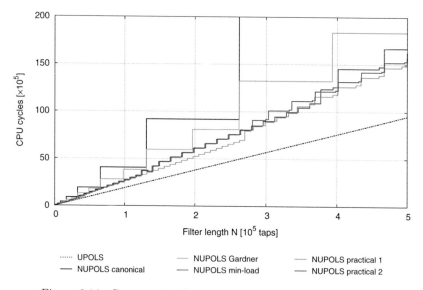

Figure 6.14.: Computational costs for the filter transformation of different convolution techniques on the test system (block length $B = 128$).

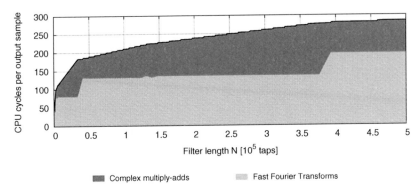

Figure 6.15.: Distribution of streaming filtering costs in NUPOLS convolution for the block length $B = 128$ on the test system.

Choice of filter partitions

The impact of the non-uniform filter partition on the computational effort is examined in the following. Figure 6.13 shows the costs for the stream filtering for different partitioned convolution methods. Therefore, a block length $B = 128$ is considered. The quasi-linear costs of uniformly-partitioned UPOLS convolution (Sec. 5.2) are marked dashed black. The non-uniform canonical partition (Sec. 6.4.5) used with standard NUPOLS convolution is the solid black curve. The lower bound is provided by the minimal-load partition (Sec. 6.5.3), marked in red. Gardner's partitioning (Sec. 6.4.6) is represented by the orange curve. The two practical partitions introduced in Sec. 6.5.4 correspond to the green (variant 1) and blue curve (variant 2).

The first observation is that Gardner's adaptions to the canonical partition lower the computational costs and do not increase it, as one might presume. The lowered number of transforms, due to the increased multiplicities, turns out to be beneficial. This example shows, that the strategy of the most rapid increase in sub filter sizes does not mark the computationally cheapest solution. The second observation is that the costs can be enormously lowered once more by the optimized minimal-load partition. It is almost twice as efficient as Gardner's partitioning scheme. However, the comparison is not fair, as minimal-load partitions lack any practicality. The practical partitions inhere more scheduling flexibility, which increases their computational costs over the lower boundary of the minimal-load partitions. Less restrictive rules (variant 2) lead to a minor increase in costs. Enforcing more freedom for the scheduling of sub convolutions (variant 1) increases the computational costs. Summing up, the filter partitioning itself marks a fundamental and important parameter for the class of non-uniformly partitioned convolution techniques. The benchmark-based and guided optimization procedure described in section 6.5.4 is very beneficial.

6.7. Complexity classes

The observed computational savings of non-uniformly partitioned convolution over uniformly partitioned convolution is now analyzed in more detail, by deriving the theoretical runtime complexities of the NUPOLS technique. Such a general examination can hardly be conducted based on algorithmicly derived partitions (like minimal-load partitions). A formal partition scheme is necessary. Here, the canonical partition was selected, as it features a simple structure and also acts as an upper bound for the computational complexity (see Fig. 6.13). Complexity classes found for this partition thus also hold for the other partitioning strategies.

Let N denote the total length of the M-segment canonical partition $\mathcal{P}_{\text{canon}} = \left([B]^1, [2B]^1, \ldots, [2^{M-1}B]^1\right)$, defined in Eq. 6.28. N is found by summation of sub filter lengths L_i (Eq. 6.27)

$$N = B \sum_{i=0}^{M-1} 2^i = B(2^M - 1) \tag{6.47}$$

Solving Eq. 6.47 for the number of segments M yields

$$M = \log_2 \left(\frac{N}{B} + 1\right) \tag{6.48}$$

For the usual case that $N \gg B$ and that B is a fixed constant, M can be approximated by

$$M = \log_2 \frac{N}{B} + \mathcal{O}(1) \tag{6.49}$$

The computational costs of each segment $[2^i B]^1$ are found by evaluating the cost function of the standard UPOLS algorithm in Eq. 5.10 with the parameters $N_i = L_i = 2^i B, P_i = 1$

$$T_{\text{UPOLS}}^{\text{stream}}(2^i B, 2^i B) = 4k \log_2 \left(2^i B\right) + 8\left(\frac{2^i B}{2^i B}\right) + 4k + 6 \tag{6.50}$$

$$= 4k\left(\log_2 B + i + 1\right) + 14 \tag{6.51}$$

This result is inserted into the accumulated costs in Eq. 6.12.

The additive term $T_{\text{ADD}}(B_i)/B_i$ is substituted with 1.

$$T_{\text{NUPOLS}}^{\text{stream}}(M) = \sum_{i=0}^{M-1} \left(T_{\text{UPOLS}}^{\text{stream}}(2^i B, 2^i B) + 1 \right)$$

$$= \sum_{i=0}^{M-1} \left(4k\big(\log_2 B + i + 1 \big) + 15 \right)$$

$$= 4k \sum_{i=0}^{M-1} i \; + \; 4kM\big(\log_2 B + 1 \big) \; + \; 15M$$

$$= 2kM(M-1) \; + \; 4kM\big(\log_2 B + 1 \big) \; + \; 15M$$

$$= 2kM^2 \; + \; 2kM \; + \; 4kM \log_2 B + 15M \tag{6.52}$$

These costs are still a function of the number of segments M. In order to relate them to the filter length N, the approximation in Eq. 6.49 is inserted

$$T_{\text{NUPOLS}}^{\text{stream}}(N) = 2k \log_2^2 \frac{N}{B} + 2k \log_2 \frac{N}{B} + {}$$

$$4k \log_2 \frac{N}{B} \log_2 B + 15 \log_2 \frac{N}{B} \tag{6.53}$$

The asymptotically dominant term in Eq. 6.53 is $\log_2^2 N/B$. This follows from $\forall x \in \mathbb{R}, x > 2 : \log_2 x > 1 \Rightarrow (\log_2 x)^2 > \log_2 x$. Again, all applied approximations have an error bounded $\mathcal{O}(1)$. Without a formal proof, the complexity class of the streaming processing of the canonical partition implemented with standard parameters yields

$$T_{\text{NUPOLS}}^{\text{stream}}(B \text{ const.}, N) \in \mathcal{O}(\log^2 N) \tag{6.54}$$

Gardner's NUPOLS algorithm [39] lies within the same complexity class. This follows from the observation, that the canonical and the Gardner partition only differ in the multiplicities of the parts. Hence, Gardner's partition can maximally double the number of spectral convolutions, but not even the overall costs. Minimal-load partitions, introduced in section 6.5.3, have even lower operation counts. Consequently, Eq. 6.54 manifests as an upper bound for all three types of partitions (canonical, Gardner and minimal-load). However, it is an open question if this bound is sharp or minimal-load partitions are asymptotically cheaper. The most important result of this section is, that non-uniform filtering in $\mathcal{O}(\log^2 N)$ (Eq. 6.54) is asymptotically faster than uniformly partitioned methods in $\mathcal{O}(N)$ (Eq. 5.12). Eq. 6.53 also unveils a different dependency of the costs on the block length B, as for prior algorithms. The common term $\log_2 N/B$ introduces a logarithmic influence of the block length. In other words, halving the block length B results in an additive component, but not in a multiplicative one. This effect can be observed in figure 6.12.

The computational costs for the filter transformation are derived similarly. Each part in the canonical partition is repeated only once ($P_i = 1$). Hence, a complete filter transformation consists of M forward FFT of the padded sub filter impulse responses. According to Eq. 5.3 the costs are given by

$$T_{\text{NUPOLS}}^{\text{ftrans}}(M) = \sum_{i=0}^{M-1} T_{\text{UPOLS}}^{\text{ftrans}}(2^i B, 2^i B) = \sum_{i=0}^{M-1} k 2^{i+1} B \log_2\left(2^{i+1} B\right)$$

$$= \sum_{i=0}^{M-1} k 2^{i+1} B\left(i + 1 + \log_2 B\right)$$

This formula is seperated and simplified using the two identities A.4 and A.6 found in the appendix

$$= 2kB \underbrace{\sum_{i=0}^{M-1} i 2^i}_{=M 2^M - 2^{M+1} + 2} + 2kB(1 + \log_2 B) \underbrace{\sum_{i=0}^{M-1} 2^i}_{=2^M - 1}$$

$$= 2kB(M 2^M - 2^{M+1} + 2) +$$
$$2kB(1 + \log_2 B)(2^M - 1) \tag{6.55}$$

Inserting the approximate $M = \log_2(N/B)$ (Eq. 6.49) yields

$$= 2kB\left(\frac{N}{B} \log_2 \frac{N}{B} - 2\frac{N}{B} + 2\right) +$$
$$2kB(1 + \log_2 B)\left(\frac{N}{B} - 1\right)$$

$$= 2kN \log_2 \frac{N}{B} - 4kN + 4kB + 2kN$$
$$- 2kB + 2kN \log_2 B - 2k \log_2 B$$

$$= 2k\left[N \log_2 \frac{N}{B} + (N-1) \log_2 B - N + B\right] \tag{6.56}$$

The major term in Eq. 6.56 is $N \log_2(N/B)$, which concludes the complexity class of a complete filter transformation for the canonical partition (without formal proof)

$$T_{\text{NUPOLS}}^{\text{ftrans}}(B \text{ const.}, N) \in \mathcal{O}(N \log N) \qquad (6.57)$$

Transforming a filter for non-uniformly partitioned convolution is hence asymptotically more expensive than for a uniform partition, which requires $\mathcal{O}(N)$ operations (Eq. 5.5).

6.8. Implementation

Uniformly partitioned convolution algorithms, reviewed in chapter 5, are conceptually simple and easy to implement. As the same operations are carried out for each processed block of the audio stream, the load distribution is even. Non-uniformly partitioned techniques in contrast, are significantly more complex and harder to implement. The challenge lies in the execution of the sub convolutions with respect to their timing dependencies. On multiprocessor machines, featuring several processors or cores, the distribution of these tasks is hardened by the necessary thread synchronization. This section discusses the implementation of non-uniformly partitioned convolution techniques and outlines recent approaches.

Scheduling

Schematic partitioning approaches, like Gardner's partitioning, result in a periodic scheduling pattern of the sub convolutions. Given a filter partition, an execution plan can be created a-priori and then hard-coded into the convolution software. In general, such approaches are reasonable, if the conditions of the processor do not vary (e.g. exclusive use, constant clock rate) and if the filtering parameters are not subject to frequent changes (e.g. fixed number of channels, sampling rate, block length and filter length). A deterministic scheduling is of primary interest in DSP solutions.

Applications in acoustic virtual reality are characterized by a time-varying filtering, with dynamically changing parameters. This does not only include the filter impulse responses themselves. Over the runtime of a real-time auralization, channels are added or removed and parts of the filter impulse responses are updated individually and with different rates. Moreover, the individual channels might be filtered with impulse responses of different lengths. The computing resources of the machine might be shared with other parts of the software (e.g. simulation methods) and are not reserved exclusively for the signal processing. In these circumstances, a static execution plan for the sub convolutions is disadvantageous and cannot adapt the changing runtime conditions.

The solution is a flexible scheduling of the sub convolutions during the runtime. Such a dynamic scheduling was first advocated by Gardner [39]. Most subsequent publications use a similar concept [71, 9, 76, 121]. The scheduling is typically realized in the following way: Each individual sub convolution is represented by a *convolution task*, which encapsulates the according channel, filter segment and additional parameters (e.g. filter update, cross-fading parameters). Tasks are created directly within the stream processing callback (interrupt), when new input blocks of data are provided. Given the availability of the input data for the next execution of a sub convolution, a new task is issued and stored in the scheduling list (*task queue*). Tasks are usually attributed with additional metadata, like the read/write positions on the input/output buffers and their execution deadlines, etc. The leading segment in the partition is computed directly within the callback. Other tasks in the list are managed by the scheduler and executed asynchronously (e.g. in individual threads). The scheduling list itself can be implemented as a queue, multi-level queue or priority queue.

The main challenge lies in the correct execution of all tasks within their deadlines. A common approach is to categorize the sub convolution tasks and treat them with different priorities. Gardner [39] presented a scheduling approach for a three-segment partition $(B^2, [2B]^2, [4B]^{1/2})$, with three different task categories: high, medium and low. All initial segments are of high priority and executed directly within the stream processing callback without interruption. This is a common approach for all non-uniformly partitioned techniques (e.g. [39, 71, 9, 76]). The other tasks (medium and low) have different priorities. In each streaming cycle a fixed number of tasks from these two categories are serviced. Low priority tasks of later segments, which require more computation, can be interrupted (*preemption*). A specialty of Gardner's methods is, that at no point in time two tasks of the same sub filter exist. In other words, new tasks of a segment are issued just in the cycle when the preceding task of this segment reached its deadline. This property makes the scheduling conceptually simple. In contrast to Gardner's scheduling, which is still relatively fix, Battenberg and Avizienis [9] realize a fully dynamic scheduler, implementing an *earliest deadline first* heuristic. This means, that tasks of later segments may receive a higher temporary priority and be boosted in their execution, given that their deadline is approaching. The scheduling is affected by further aspects which are discussed in the following.

Preemption

A challenge of non-uniformly partitioned convolution on single-core processors is to ensure that sub convolutions of larger granularity do not block the execution flow of higher priority tasks. For long filters, the largest segments in a partition can consume a significant computation time. A single FFT or IFFT can compute relatively long in comparison to the available time budget or even exceed it. A wrong scheduling decision can cause priority inversion for the tasks, violently disturb the sensitive execution. In these cases manual preemption of the sub convolution tasks is strictly advised to keep the individual work packages at a fine granularity. The term 'manual' means breaking down the sub convolutions into their computation steps *by hand* and not by the operating system (OS). This requires a significant programming effort, but is reported to improve the performance and stability significantly [9]. The execution of a sub convolution can be suspended after each individual operation (FFT, spectral convolutions, IFFT, etc.) Simple operations, like spectral convolutions and accumulations, can easily be decomposed further. With respect to the distribution of costs (see Sec. 6.6), this already allows breaking a large amount of computation into easier manageable units.

Subdivision of transforms

Fast transforms are usually computed using high performance libraries, which execute the full transformation at once. They do not allow an interrupted computation of FFTs and IFFTs, which can be suspended and finished later on. A solution to this problem is a manual decomposition of the transforms. Hurchalla [47] presented, how a large FFT respectively IFFT is decomposed into a sequence of short length transforms, using the decimation-in-frequency (DIF) scheme (see Sec. 2.5.2). These shorter transforms can be computed block-by-block, delivering sub sets of the DFT coefficients, which can be immediately processed. The technique is referred to as '*time-distributed FFT*'. It allows a nearly constant workload [9] over the individual cycles. Hurchalla [47] points out, that the computational overhead is relatively little. Examinations by Battenberg and Avizienis [9] indicate a significant benefit of the technique in single-core implementations.

Multi-processor systems

Bulky sub convolutions are usually much less of a problem on multi-processor systems, where several processing threads serve the sub convolutions. Even if one of the worker threads is occupied longer over the duration of multiple stream cycles, the execution of the other tasks does not stagnate in the meantime and continues in parallel. The difficulties on multi-processor systems lie in the inter process communication between the different worker threads. A detailed discussion of the implementational aspects of non-uniformly partitioned convolution on multi-processor systems is found in [9]. Battenberg and Avizienis [9] consider multi-core implementations on standard PCs and address the sub convolution scheduling on conventional operating systems. Here, the thread synchronization mechanisms themselves consume a significant runtime. Getting it wrong, can cause severe losses in the computational efficiency [9]. *Lock contention*, is a particular issue on multi-core implementations with a multitude of worker threads, when conventional synchronization concepts are used (e.g. thread conditions, mutexes, semaphores, etc.) The limitations can be overcome using atomic operations [9]. A further question is, how to assign the individual sub convolutions to the available processors or cores. Battenberg and Avizienis [9] examined two strategies for multi-channel filters: Each core processes an individual channel and each core is assigned a specific segment and processes all channels for this segment. They report, that the latter strategy performs significantly better.

6.9. Filters with multiple inputs and outputs

The regarded non-uniformly partitioned methods can easily be extended to serve multiple inputs and outputs, as it was possible for UPOLA/UPOLS MIMO filters (see section 5.5). This is due to the principle, that a non-uniformly partitioned convolution is assembled from uniformly-partitioned sub filters (see figure 6.1). Several applications for large-scale MIMO FIR filters can be shown up: Considering time-invariant filtering, they are an essential tool for room acoustic equalization and the electro-acoustical enhancement of the reverberation in a room [79]. Such systems implement an FIR filtering matrix. M microphones are placed within the room of interest. Their signals $x_i(n)$ are filtered and artificially enhanced with additional reverberation. The resulting cues $y_i(n)$ are played back into the room using N loudspeakers. In the most general manner, each of the $M \cdot N$ filtering paths between a microphone and loudspeaker can have an individual filter impulse response $h_{i \rightarrow j}(n)$. As before, a combined approach for such filters is appealing. In real-time room acoustic auralization [64, 90] they have an application for time-varying MIMO FIR filtering. Binaural rendering techniques generate a two-channel signal from a mono signal. Computation can be saved here, by transforming the signals of each virtual sound source only once. In contrast to uniformly-partitioned approaches, multiple FFTs of different transform sizes are computed from the same input samples. Several sound paths can superpose at a virtual listener (receiver). Here, a potential improvement is the summation in the frequency-domain. This motivation gets greatly emphasized for auralizations with sound transmission [119]. Under these circumstances, not one, but multiple filtering pathes can exist between a virtual source and receiver. This encourages the research of specialized MIMO filter for these applications.

A non-uniformly partitioned MIMO filter typically utilizes a common filter partition for all of its filter paths $h_{i \to j}(n)$, similar to multi-channel non-uniformly partitioned convolutions [39, 9]. Advantageously, this also implies a common scheduling pattern for all of the sub filter convolutions. Furthermore, this increases the possibilities of reusing transform input spectra (e.g. matching transform sizes). The use of different partitions would be reasonable, if the individual filters $h_{i \to j}(n)$ require different filter update parameters (cp. section 6.11). However, for the above stated reasons, this complicates the problem.

Let it be assumed that all filters $h_{i \to j}(n)$ use the same filter partition $\mathcal{P}$. In this case, a unified implementation can realize individual input buffers (as in figure 6.1) for each input and reuse all computed DFT spectra (of different granularities K_i) for all connected uniformly-partitioned segments. Thereby the number of input FFTs is reduced and so are the computational costs.

In uniformly-partitioned methods, the superposition of multiple filter paths at outputs can be efficiently realized in the frequency-domain as well. This is possible due to the fact, that all sub filters make use of the same transform size K. A similar approach is not directly applicable for the outputs in a non-uniformly partitioned MIMO filter, as here the individual sub filters use different transform sizes K_i. Consequently, their results need to be overlap-added in the time-domain (see figure 6.1). Fortunately, an accumulation of the sub filter results in the frequency-domain can be realized *segment-wise*, before the final overlap-add in the cumulative output buffer. This requires a synchronous scheduling of sub convolution for all sub filters of all filter paths $h_{i \to j}(n)$.

The resulting non-uniform MIMO convolution algorithm is briefly outlined: Each input is attributed with an independent input buffer for storing and repacking the input samples. It is combined with a transformer, that provides the input spectra in the required granularities K_i. Each output is implemented with the frequency-domain accumulation buffers and inverse transforms for each segment of the partition (see figure 5.2), followed by the final time-domain overlap-add step of the non-uniform convolution (see figure 6.1). Superposing sub filter results are accumulated at the output-side for each segment. Then, each filter path $h_{i \to j}(n)$ between an input and an output only requires to perform a series of partitioned spectral convolutions, which are controlled in the usual way, using sub convolution tasks.

Also in non-uniformly partitioned techniques, a large share of the computation falls back to fast transforms. This indicates the benefit of the proposed MIMO technique. A detailed examination is skipped here, due to the complexity of parameters (e.g. numbers of inputs and outputs, filter partitioning, latencies, etc.).

6.10. Filter networks

For real-time filtering with long FIR filters, non-uniformly partitioned convolution became the primary tool. Some applications demand assemblies of these already complex FIR filters. Again, an example is found in acoustic virtual reality: the real-time auralization of buildings including the simulation of sound transmission. Such scenarios are very complex and typically tackled using a discretization in space (portal paradigm) [104, 119], which limits the number of sound propagation paths to a manageable, finite number. Wefers and Schröder [119] describe the occurring sound paths between sources and receivers in form of a directed acyclic graph (DAG). Using a rule set, this DAG is transformed into the corresponding filtering network, that realizes the audio rendering. Sections of air-borne sound propagation are thereby described by RIRs. Sequential topologies (e.g. room-portal-room) result in cascades of filters. Several parallel propagation paths join in the interconnection joints (portals). The technique presented in [119] first merges these networks into their equivalent impulse responses (facilitating Eq. 4.20 and Eq. 4.26 in section 4.3) and then performs the audio rendering, using conventional single-input single-output (SISO) non-uniformly partitioned convolution. A unified frequency-domain approach for these networks would be desirable, but is hindered by several problems. These are outlined in the following paragraphs. The problem of sequential or parallel assemblies of non-uniformly partitioned convolutions seems not to be considered in the literature so far. This section regards assemblies of these filters and discusses potential solutions and pitfalls.

Uniformly-partitioned frequency-domain filters, like UPOLA and UPOLS (see chapter 5), use the same block length and transform size for all of their sub filters. This makes it possible to connect them directly the frequency-domain (see section 5.6). In contrast, non-uniformly partitioned convolution processes the sub filters with different block lengths and transform sizes. The spectra are not compatible among the sub filters, as they have different sizes. This requires to accumulate the results of all sub filters in the time-domain, as shown in figure 6.1.

An example of a parallel assembly of partitioned filters is shown in figure 6.16(a). For simplicity, the block diagram is written in the time-domain. Note, that both filters (dotted square) share a common partitioning. All

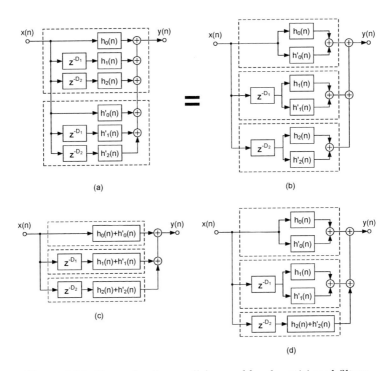

Figure 6.16.: Example of a parallel assembly of partitioned filters

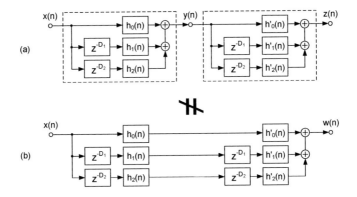

Figure 6.17.: Example of a sequential assembly of partitioned filters

sub filters $h_i(n)$ and $h_i'(n)$ have the same lengths and equal branch delays D_1, D_2. The composite structure rearranged, by joining all sub filters of common delays (figure 6.16(b)). This layout consists of parallel uniformly partitioned filters, which can be effectively realized in the frequency-domain (see section 4.3). A further simplification is achieved by adding all parallel sub filter impulse responses $h_i(n), h_i'(n)$ (see figure 6.16(c)). This reduces the number of spectrum multiplications and accumulations in the filter segments. Depending on the application, this can also be done for a subset of the segments, as illustrated in figure 6.16(d).

A sequential arrangement of filters is more difficult: Figure 6.17(a) illustrates an example of a cascade of non-uniformly partitioned filters. The essential point in this block diagram is, that each sub filter $h_i(n)$ of the first filter contributes to the intermediate signal $y(n)$. Hence, this point of superposition $y(n)$ can not simply be removed, as done in figure 6.17(b). In essence, each sub filter of the first partitioned filter influences the input of each sub filter of the second partitioned filter. Consequently, both block diagrams in figure 6.17 (top and bottom) are not equivalent. The realization of cascades of NUPOLS or non-uniformly partitioned Overlap-Add (NUPOLA) filters depends on an computationally efficient method to the accumulation of sub filter results (as for $y(n)$). The challenge lies in the fact that the samples are accumulated and provided in different granularities (transform sizes). Gardner [39] pointed out the possibilities of computing a larger DFT spectrum from already known shorter DFT spectra of blocks neighboring in time. Given, that all accumulations could be carried out on DFT spectra of the smallest size, his approach could be used to accelerate the redistribution after the accumulation. Still, the large granularity DFT spectra would have to be decomposed into sequences of smaller DFT spectra first. It remains an open scientific question, if such an approach can be realized with computational benefit.

Summing up, the implementation of networks of non-uniformly partitioned filters in the frequency-domain is partially possible. A mandatory requirement is a common filter partition. Parallel assemblies of non-uniformly partitioned frequency-domain filters can be realized conceptually simple, without only minor adaptions. This matches the results for unpartitioned and uniformly-partitioned filters (see sections 4.3 and 5.6). For sequential arrangements it is hard to obtain a computational benefit from a frequency-domain implementation. The pivot point is identified in the efficient realization of the sub filter accumulations, as discussed above. It might be a better option to replace a cascade of NUPOLS or NUPOLA filters by an equivalent single, combined filter, as discussed in section 4.3. Aware of the extent, a detailed examination of the computational costs is not carried out here. The proposed implementation strategies should be carefully examined for a given application.

6.11. Filter exchange strategies

Müller-Tomfelde [72] was presumably the first who addressed the exchange of filters in non-uniformly partitioned convolution algorithms. The findings of his paper are briefly reproduced in the following. Generally, the filter exchange is subject to a trade-off between a coherent transition and the response time (exchange latency). Müller-Tomfelde [72] introduced two strategies which align with these two possibilities. Figure 6.18 illustrates both strategies in four examples, in style of [72]. A 4-segment Gardner partition is considered. The time axis has a grid of length-B blocks. Black dots denote triggers for the exchange, dark gray hatched areas correspond to crossfade regions, light gray backgrounds mark intervals, where both filters (old and new) are computed in parallel.

Example 6.18(a) shows a coherent filter exchange. The filter exchange can be triggered for all segments instantaneously. Fortunately, all segments compute in the next cycle and the transition spans over a single cycle. This example marks the best case, as the response and transition time are little. In example 6.18(b) the same filter exchange is trigger one cycle later in time. Unfortunately, the next coherent execution of all segments happens eight cycles later and thus the response time is maximally long, marking the worst-case. The individual segments must be triggered at later points in time in order to achieve a coherent transition. For the strategy of a coherent exchange, the transistion time is constant and can be adjusted by the crossfade length, whereas the response time can fluctuate and depend on the actual trigger time.

Examples 6.18(c,d) illustrate an asynchronous filter exchange. For this strategy all segments are always triggered immediately. In 6.18(c) the exchange takes place in-order, in other words it moves along the impulse response. Nevertheless, the complete transition is not continuous and in-between it starves (light gray). This can be overcome by choosing larger crossfades for the longer segments. In comparison to the examples 6.18(a,b), the response time is low (only one cycle), whereas the transition time is longer. In example 6.18(d) the same exchange is triggered two cycles later in time. Here, the consequence is a slightly reduced transition time. It can be summarized, that an asynchronous filter exchange results in a constant response time, but an unpredictable transition time, which can in the worst-case be as long as the longest segment.

These examples considered the Gardner partitioning (see section 6.4.6), which is characterized by a periodic execution pattern of the sub convolutions. Other partitioning schemes do not necessarily have this property. This can affect the filter exchange capabilities as well. Following the general theoretical considerations in section 6.4.3, the use of power-of-two sub filter

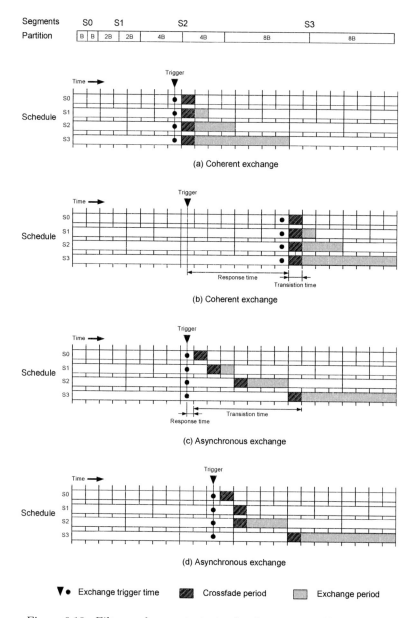

Figure 6.18.: Filter exchange strategies for time-varying filtering with non-uniformly partitioned convolution, as proposed by Müller-Tomfelde [72]

lengths L_i is advised for a coherent filter exchange. In this case the points in time, where all segments execute in parallel are narrowly spaced (small least-common-divisor). A worst-case choice mark relatively prime sub filter lengths L_i (large least-common-divisor), resulting in very long worst-case response times. These considerations make clear, that also the filter exchange should be considered in the optimization process of an impulse response.

6.12. Real-time room acoustic auralization

This section discusses the use of non-uniformly partitioned convolution algorithms for the real-time audio rendering with room impulse responses (RIRs). The considerations are based on the following example case: A reverberation time of 2.0 s , a sampling rate of 44100 Hz and a streaming block length of $B = 128$ samples. The destination filter length is $N = 88200$ taps. This example matches typical parameters in auditory environments (e.g. of a concert hall). Again, the performance is studied for the test system. Let it be assumed, that the early reflections can be computed using the image source model in 30 ms and that a ray-tracing procedure for the diffuse decay consumes 500 ms in total. The throughput (simulations per second) for the image source filter is 33.3 Hz and 0.5 Hz for the ray-tracing. For comparison, a uniformly-partitioned convolution is regarded. The standard UPOLS algorithm partitions the filter into 690 length-128 sub filters. Each filtered output sample consumes 2018 CPU cycles. A filter transformation of the entire impulse response takes approximately $700\,\mu s$. The advantage of the algorithm is its uniform load distribution. A disadvantage is that the entire computation falls into the time critical audio callback. The filter update rate of 345 Hz widely exceeds the throughput of the GA simulations.

The corresponding non-uniform minimal-load partition is $\mathcal{P}_{\mathrm{opt}} = (\, 128^7\, 1024^7\, 8192^{10}\,)$, covering 89984 taps. The NUPOLS algorithm, operated with this partition, has an total cost of 204.1 CPU cycles per filtered sample. This is $9.9\times$ more efficient than uniform convolution. However, all segments in this partition have zero clearance, resulting in a highly unfavorable load distribution of the algorithm. An alternative partition with suitable scheduling flexibility is needed. The optimized partition of strategy 1 (see Sec. 6.5.4) is $\mathcal{P} = (\, 128^2\, 256^4\, 1024^8\, 8192^{10}\,)$. The partition is visualized in figure 6.19. It does not differ too much from the minimal-load partition. The design principle lowered the multiplicity of the leading segment, which shifts away the computation from the time-critical audio callback into the background threads. This requires to use a fourth intermediate segment 256^4. The remaining partition is rather similar to the minimal-load solution. Operating the NUPOLS algorithm with $\mathcal{P}$ requires 240.4 CPU cycles per sample. This increases the costs by 18%, but is still $8.4\times$ more efficient than a uniform partition.

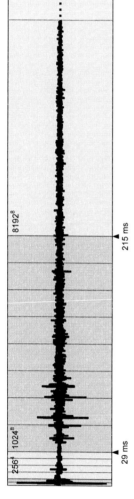

Figure 6.19.: Practical non-uniform filter partition for a filter length of 88200 taps

Segment	Time span [cycles]	Cost stream [CPU cycles/ sample]		Filter transformation runtime	Filter update rate	Filter exchange latency
128^2	1	45.8	(19%)	2.0 μs	344.5 Hz	2.9 ms
256^4	2	48.0	(20%)	7.7 μs	172.3 Hz	5.6 ms
1024^8	4	61.1	(25%)	69.3 μs	43.1 Hz	23.2 ms
8192^{10}	12	85.5	(36%)	950 μs	5.4 Hz	186 ms

Table 6.1.: Properties of the non-uniform filter partition ($128^2\,256^4\,1024^8\,8192^{10}$)

Table 6.1 lists the properties of $\mathcal{P}$ and its segments in detail. The partition allows a flexible scheduling. The segments 2-4 have to be computed within time spans of two, four respectively twelve cycles. Only 19% of the computation (first segment) are spent in the audio callback.

An entire filter transformation demands about 1030 μs of computation time, about 50% more than for a uniform partitioning. However, the examination of each segment reveals, that the individual transformation runtimes are negligibly short compared to the simulation times. The filter updates exceed the simulation throughput. This allows to implement a coherent filter exchange, with respect to the different parts of the impulse response. The direct sound (HRIR) aligns with the leading segment and should be exchanged as frequently as possible. The time span of the early reflections, roughly the first 50-80 ms after the direct sound impulse [116], corresponds to the first 2000-3500 taps at the given sampling rate. Thus early reflections are associated with the segments 256^4 and 1024^8. These span over time range of 29-215 ms in the impulse response, which even exceeds the range 50-80 ms assumed above. The last segment 1024^8 covers all remaining filter coefficients after 215 ms . In a real-time auralization, the parts of the impulse response are simulated asynchronously in parallel [64, 89, 120]. The given partitions allow to update the direct sound filter coefficients with 344.5 Hz and a latency of $\leq 2.9\,ms$. The rate for early reflections is determined by the third segment with a block length of $B_2 = 1024$. It allows updating the filter with 43.1 Hz. For the update of the early reflection filter a coherent filter exchange is advised, keeping the transition time short. The latency of a coherent exchange is bound by 23.2 ms . The diffuse reverberation tail of the impulse response is dealt with separately. It falls within the segments three and four. Here the maximal update rate is 5.4 Hz resulting from the block length $B_3 = 8192$. A coherent exchange of the segments three and four would result in worst-case exchange latency of 186 ms , which is significant. This suggests to update the diffuse decay part asynchronously.

The considered example impressively illustrates the enormous speedup of a non-uniform partitioning. However, it also illustrates how various aspects influence the optimal choice of partition. Unfortunately, only little is known about meaningful parameters for real-time auralization, in particular the perception and audibility of the filter exchange strategies. These aspects remain an important topic for future research.

6.13. Perceptual convolution

All previously discussed convolution algorithms are exact, apart from floating-point round-off errors. They compute the output signal of a linear filter precisely, without approximations. For several applications it is questionable if such a high level of accuracy is actually needed, or if a less accurate filtering allows a reduction of the computational costs. This particularly applies for room acoustics and artificial reverberation. Here, parts of the simulation algorithms are non-deterministic and based on stochastic processes. Recent research indicates, that the simulation effort can be lowered by a surprising degree, without affecting the perceived quality in virtual environments [88]. The properties of human hearing should be considered in the convolution algorithm as well.

Recently, perceptual measures were incorporated into fast convolutions for the sake of a reduced computational complexity of the filtering. Lee et al. [61] proposed a novel technique for artificial reverberation using FIR filters, referred to as *fast perceptual convolution*. Their approach exploits the frequency-dependency of the reverberation time in room acoustics [60]. In typical rooms, the high-frequency energy of the sound field decays faster (e.g. due to frequency-dependent absorption). Higher frequencies are attenuated stronger, which results in shorter reverberation times than for lower frequencies. A uniform filter partition of a RIR can be seen as a discrete-time short-time Fourier transform (STFT) without overlap and rectangular window. For RIRs it can be observed, that higher frequencies in DFT spectra of later sub filters get more and more attenuated, the later the sub filter is placed.

Lee et al. regard partitioned frequency-domain convolution techniques with a uniform partitioning of an RIR (UPOLA, UPOLS) and explicitly consider an offline processing with large block lengths (e.g. $B = 4096$). Their technique *truncates* the spectral convolutions of sub filters, by only considering DFT coefficients with significant energies above the threshold of hearing. This significantly reduces the number of complex-valued multiplications and accumulations, which in uniformly-partitioned algorithms makes up for a large part of the computation. In their examples and listening tests the authors drop 60% of the spectral convolutions without a loss in perceived audio quality. They state speed up of the overall RIR convolution in the range of 30%. Primavera et al. [76] applied the concept by Lee et al. [61] to non-uniformly partitioned convolution for real-time filtering. They examined both techniques, UPOLS and NUPOLS, with three room impulse responses of different reverberation times. For a uniform partitioning they observed slightly larger speedups than those reported by Lee et al. [61]. For a non-uniform partitioning, the achieved total speedup due to psychoacoustic enhancements reached up to approximately 30%. Their results also showed the tendency, that the savings drop for the cases of small block length (lower latency) and long room impulse responses.

Perceptual convolution is a promising concept, allowing for significant speedups of the FIR filtering in room acoustic auralization. However, the methods are relatively new and in an early stage of research. An issue of the originally proposed fast perceptual convolution techniques [61, 76] is that the truncation thresholds are based on the filter impulse responses only. For the application in virtual environments with room acoustic, it is important to consider the magnitudes of the input and resulting output DFT coefficients as well. The actual perceptual thresholds would presumably allow for an even larger reduction of complex-multiplications. In order to consider absolute thresholds of hearing, the sound pressure levels at the listener's ears must be considered, which asks for a fully calibrated auralization system with a correct representation of levels.

6.14. Summary

This chapter considered fast algorithms for computing running convolutions by partitioning the filter impulse into sub filters of varying lengths. These non-uniformly partitioned techniques are motivated by the observation, that the computational costs of the filtering are reduced by selecting larger block lengths. Given a causal partition of the impulse response, the sub filters can be realized with growing block lengths, without sacrificing a low input-to-output latency. This allows a massive reduction of the computational costs, even if the number of fast transforms increases.

The construction of filter partitions is subject to conditions. In order to preserve the real-time compliance of the convolution, the sub filter length cannot increase too fast. Their growth is limited by causality conditions. Gardner [39] was mistaken in his assumption, that a most rapid increase in sub filter sizes would lead to the minimal computational costs. His partitioning scheme does not exploit the potential of frequency-domain delay-lines (FDLs) within the uniformly partitioned filter segments of a partition. The minimal computational costs, achieved by model-based optimization of the partition, are almost twice as efficient as Gardner's approach. On the contrary, they are not of practical relevance and serve as an interesting theoretical bound only. The runtime complexity lies within the class $\mathcal{O}(\log^2 N)$, making non-uniformly partitioned convolution techniques asymptotically faster than uniformly-partitioned ones in $\mathcal{O}(N)$.

The choice of partition is not trivial and influenced by several aspects. The issue of minimal-load partitions are their critical timing dependencies. Even if they have the least accumulated costs (numbers of operations), the computational is very poorly distributed over the runtime. The cause of this problem is absence of flexibility in the execution of the sub filter convolutions, which concentrates their computation at distinct points in time. The consequence are load peaks, which eventually become a bottleneck for the real-time processing. Less rigid timing dependencies allow the sub convolutions to be shifted along the time, leveling out the load distribution. A certain minimal flexibility can be enforced during the optimization. More scheduling flexibility increases the computational costs. Still, a large computational benefit is gained over Gardner's method by specific optimization of the partitioning. Many aspects must be considered thereby, including the computing platform (single-core vs. multi-core), the underlying implementation (preempted and/or time-distributed), the required latency and filter exchange parameters.

The concepts of uniformly-partitioned MIMO filters apply to non-uniformly partitioned convolution as well. This is reasoned by the fact, that a non-uniform convolution is composed from uniform sub convolutions. The same accords for parallel assemblies of non-uniform filters. Cascades however, are difficult to implement. A time-varying filtering is once more influenced by the filter partition. The two opposing factors are response time versus transition time. The larger block lengths involved in the algorithms, effectively lower the achievable filter update rates. Recently, perceptive measures have been incorporated into convolution algorithms used for artificial reverberation. These techniques allow a further reduction of the computational costs. What meaningful quality parameters are for this type of FIR filtering (e.g. parameters for a time-varying filters, perceptional thresholds) has not been examined in depth and remains a topic for future research. This also holds for the implementation of the techniques on different hardware and software platforms. On standard PCs they allow an impressive performance, building a foundation for the real-time auralization with comprehensive simulations of the room acoustic.

7. Benchmarks

A central aspect of this thesis is the research of convolution algorithms with respect to their computational complexity. The different algorithms are characterized by a variety of parameters (e.g. block lengths, filter lengths, transform sizes, filter partitions, etc.). Subject of the research is not only how to select these parameters in an adequate way, i.e. minimizing the computational load, but also to understand their dependencies, in order to draw general conclusions. A purely theoretical analysis of the algorithms rarely provides objective statements on the actual performance of an algorithm on a machine.

This chapter introduces a benchmark-based approach to model the computational complexity of convolution algorithms with respect to the target hardware. A dedicated test system is introduced, on which all examinations were conducted. The requirements for the benchmark method are outlined and the measurement procedure is explained in detail. Fundamental arithmetic operations and different discrete transforms were measured and compared in their execution speeds. The validity and accuracy of the techniques is verified. The relations between the computational costs in theory and practice are analyzed. Special attention is paid to the computation of the fast Fourier transform for arbitrary sizes. It is shown, that transforms of other sizes than powers of two can be computed efficiently, making them worth the consideration in the convolution algorithms.

7.1. Benchmark-based cost models

The formal description of an algorithm, its runtime and space complexity, does rarely provide an accurate statement on its performance on an actual machine. The number of required instructions or clock cycles on a processor is not only determined by the overall number of arithmetic operations of an algorithm. Early hardware, for instance, executed multiplications several times slower than additions. As a consequence, these operations were often distinguished in the analysis of algorithms. In many cases, algorithms have been optimized mainly with respect to the number of multiplications. Improvements in hardware design evened out this imbalance between the basic arithmetic operations, which nowadays have similar, if not equal costs. With the introduction of the cache hierarchy, the memory requirements of an algorithm became more and more important and the data sizes could heavily impact the performance. As long as the data fitted the caches, the computation was fast. After exhaustion of the cache limits the execution speed tumbled. Advances like pipelining, which came along with branch prediction, further contributed the discrepancy that an algorithm performs differently on paper than on an actual machine. The practical comparison of algorithms and implementation variants, requires qualitative statements. Unfortunately, the resulting performance of an algorithm on a target machine is hard to derive from the theoretical descriptions alone. In spite of these difficulties, it is necessary to appropriately assess the computational performance of signal processing algorithms on practical hardware. Actual runtimes, clock cycles or numbers of instructions on a target machine must be considered. A simple and accurate solution to the problem is to benchmark (measure) an algorithm and its variants for parameters of interest on the target machine. Unfortunately, the vast number of feasible parameter combinations makes a comprehensive study often unhandable.

This thesis introduces a hybrid semi-empirical approach, *benchmark-based models*, which combines basic measurements with analytic models of the computational costs. The foundation are measurements (benchmarks) of basic operations and functional building blocks on a target system. Based on these measures, cost models (mathematical formulas) are defined, that allow predicting the complexity of algorithms, that are assembled from these basic operations. Reasoned by the influences discussed above, this can only work to a certain degree. An essential requirement are representative benchmarks. Many signal processing algorithms, including fast convolution algorithms, follow the programming paradigm of *stream processing*. They are characterized by a high density of arithmetic operations, a relatively uniform memory access pattern and comparatively few context-sensitive branch decisions. In contrast, tree-based search algorithms (e.g. binary space partitioning in Geometrical Acoustics), have few arithmetic operations and highly context-dependent branching, making them difficult to model, using the presented approach.

All practical examinations in this thesis were done on a dedicated test system. Its hardware and software configuration is introduced in the following.

7.2. Test system

Hardware

The test system is a standard desktop computer with an *Intel Core2 Quad Q6600* quad-core processor, clocked at 2405 MHz. This CPU supports the SSE1, SSE2 and SSE3 instruction sets for vectorization. Each of the four

CPU cores has 256 kB of exclusive level-2 cache available. Each processor (4 cores on a die) has shared access to 8 MB of level-3 cache. Both CPUs are interconnected using a QPI bus. The system is equipped with 4 GB of DDR3 RAM.

Software

The operating system is Microsoft Windows 7 SP1 (64-Bit). The code is written in C++ und built using Microsoft Visual C++ 2010 SP1. Release mode binaries are generated with the following compiler settings: complete optimization, full function inlining, 16-Byte memory alignment and fast floating-point arithmetic. 32-Bit floating points are commonly used as the data type for all considered operations. Basic arithmetic operations (e.g. complex-valued multiplications) are implemented by the help of the Intel Performance Primitives (IPP) library version 7.0, utilizing the vectorization capabilities of the processor. Discrete transforms (including the FFT, DST, DCT and DHTs) are computed using FFTW version 3.3.3. For the sake of scientific research, the FFTW library is favored for several reasons: Its design and internal structure is well-documented [36] and its source code is available to the public [34]. It supports widest range of discrete transforms (including DHTs and DTTs). Moreover its internal state can be inspected, like the actual implementation strategy for a given FFT.

7.3. Benchmark technique

As mentioned afore, the aim is to predict or forecast the resulting computational complexity of a convolution algorithm on a target machine. If this prediction is sufficiently accurate, the properties of the algorithm can be studied on by a semi-empirical runtime model of the algorithm. Absolute measures (CPU cycles or runtimes) are primarily interesting when comparing different machines or different high performance libraries. For the optimization of convolution algorithms, the focus also lies on the cost relations between different operations. With respect to fast Fourier transform (FFT)-based fast convolution, the computational costs of FFT transforms versus complex-valued multiplications and additions are most important. If, for instance, an FFT is comparably cheap to compute with respect to a complex-valued multiplication, the optimal convolution strategy might favor more FFTs, saving multiplications and vice versa.

Two types of benchmarks were conducted for the research in the thesis. Firstly, measurements of the runtimes of basic arithmetic operations (e.g. complex-valued multiplys) and discrete transforms (e.g. FFT variants).

These are the building blocks for the convolution algorithms. Secondly, measurements of the runtimes of complete convolution algorithms assembled from these basic operations. These measurements were used to validate the semi-empirical cost models. The benchmark requirements, the measurement procedure and the storage of results are presented in the following.

7.3.1. Performance profiles

A complete benchmark consists of a series of single measurements, which spread along several operations and different problem sizes of interest. Each single measurement considers a specific operation (e.g. an in-place real-data FFT) with a specific problem size (e.g. the transform size). Depending on the extent of the benchmark (number of operations, range of sizes), the execution of a complete series of measurements can take several minutes to several days. The catalog of measurement results forms a *performance profile* of the target system. This profile is the foundation for the evaluation of cost models and the optimization of convolution algorithms.

7.3.2. Measurement procedure

The validity, accuracy and robustness of the benchmarks is of great importance. Ideally, basic operations should behave similarly when measured individually or when they are integrated in a convolution algorithm. As shown later, this is only possible to a certain degree, limiting the accuracy of the cost model. Running a benchmark at different points in time (morning, evening, next week) shall deliver the same results, given that the system is in the same state (e.g. idling). Simply measuring a single operation just once is not viable. Upon the very first function call, it is very likely, that the required data is not in cache and the recorded runtime is not representative. Therefore, series of measurements are recorded and average values are calculated from these.

Collective measurements

A common and widely-used approach is to measure the operation multiple consecutive times in a loop, as defined in algorithm 2. The runtime of the operation under test is then approximated by dividing the cumulative runtime by the number of loop iterations. Typically, a fixed number of measurements in the beginning is discarded for the aforementioned reasons. After these *preruns* the runtimes converge and become stable.

$$t_1 = \text{getTime}();$$
for $i = 1$ **to** N **do**
$\quad$ runOperation();
end
$$t_2 = \text{getTime}();$$
$$t = (t_2 - t_1)/N$$

Algorithm 2: Collective measurement

$t =$ array 1 **to** N
for $i = 1$ **to** N **do**
$\quad t_1 = \text{getTime}();$
$\quad$ runOperation();
$\quad t_2 = \text{getTime}();$
$\quad t(i) = (t_2 - t_1)$
end

Algorithm 3: Individual measurement

Individual measurements

In benchmarks with a fixed number of measurement iterations, the duration of the measurement depends on the complexity of the operation under test. A short-length FFT might compute within 100 ns . So a 1000 iterations span over approximately 100 μs . In contrast, a single long-length FFT can require 100 μs and the iterations here span over 100 ms . As the benchmark program executes a large number of single measurements consecutively, it will not be granted exclusive use of the processor. Depending on the scheduling quantum of a multithreading operation system, the CPU is taken away from the benchmark process every now and then. This happens also on multi-core and multiprocessor machines and there is no way to prevent this. In the meantime the execution of the benchmark is halted, while the time measurement continued. Consequently, the collective runtime gets polluted with false measurement values, due to thread context changes. The values of these outliers strongly differ from the mean values and can cause significant deviations in the final result. It was observed, that the number of mismeasurements affects longer operations more than shorter operations. Also measurements of short operations are affected by the problem. But due to the longer execution time of whole loop, such measurements are more exposed to thread changes. Consequently, longer running operations are more affected by these outliers.

In order to overcome this problem each operation is measured individually using algorithm 3. Each recorded runtime $t(i)$ is stored in an array. An advantage is that this enabled post-processing of the sequence and its individual samples. Outliers can be detected and removed from the sequence. A downside is, that individual measurements demand an even more accurate timing, than collective measurements. This asks for very high resolution timers, which are regarded in the succeeding section. The method is calibrated by dropping the function call of the operation under test and just measuring the start-stop times. This *offset* is later subtracted from every measured time. In order to increase the accuracy for measurements of short operations, without prolonging the execution of the full benchmark, a fixed measurement duration of 100 ms was chosen. Blocks of individual measurements are executed and concatenated until this time budget is exhausted.

7.3.3. High resolution timing

Small operations, like adding two length-64 vectors or computing an 64-point FFT, compute within a low number of processor cycles. At modern clock rates, this results in very short runtimes on the range of several ten to a few hundred nanoseconds. Measuring such operations accurately demands timing with the highest precision. Even high resolution timers provided by today's operating systems proved not to be sufficient. On the test system for instance, the *QueryPerformanceCounter* function delivered an effective resolution of 425 ns (2.35 MHz timer frequency).

The most precise time source available are time-stamp counters (TSCs) in the processor cores. A TSC is a 64-bit integer register that is incremented every CPU cycle. These registers must be used with precautions. Out-of-order processors can reorder the stream of instructions for faster execution. Timer queries must be strictly prevented from being shifted in the code. This was achieved using an enforced in-order *CPUID* instruction. Each core has an individual time-stamp counter; time-stamp counters along cores are not synchronized. The benchmark was fixed to a single core. Modern processors adjust their frequency by demand (CPU frequency scaling). In order to measure *time* with the TSC, such clock changes must not occur during a measurement, otherwise the results are faulty. Frequency scaling was deactivated during the execution of the benchmark. The TSC timing delivers the consumed number of CPU cycles. Dividing this number by the CPU frequency results in the runtime. For the sake of comparing practical measures with theoretic results, CPU cycles are more convenient than runtimes, which depend on the processor's clock rate.

7.3.4. Initial behavior

In a sequence of consecutive measurements, even when executed on the same data, the first couple of samples do not reflect the transient behavior of the operation of interest. Especially the very first measurement has a runtime much higher than the eventual average value. This can have several causes: Mainly, the data on which the operations are computed might not be cached yet and cache-misses prolong the execution. The branch prediction of the processor might not yet be adapted to the executed code of the operation. Also the I/O characteristics of the benchmark process might change when the actual measurement starts and the OS needs some time to adapt the process scheduling.

By inspection of a large variety of different operations and problem sizes, it was revealed that the runtime converges close to the average runtime value after 2-3 measurement iterations. In face of the fact, that several 10,000 to 100,000 individual measurements are performed for the total measurement of one operation and size, the fixed number of ten preruns was selected as a safe and reasonable number. From each measurement performed, the first ten samples are discarded and not considered in any later statistic evaluation.

7.3.5. Transient behavior

For a fair comparison, all operations must be measured by equal means, regardless if they have very small or long runtimes. A challenge for small operations is the necessary timer resolution (see section 7.3.3). Benchmarks of longer operations become affected by the OS scheduling. In the worst-case, the OS takes the CPU away from the benchmark thread during a measurement. The benchmark thread will regain the CPU a significant time later. A consequence is a wrong, heavily increased measurement value. These outliers must be filtered from the sequence of measurements. The probability that an individual measurement gets interrupted by a thread context change increases with the computation time of the operation under test. On a multithreading OS such effects can not be prevented, even when the system contains several independent CPUs or cores and the benchmark is executed with real-time priority. The target system had 42 processes and 552 active threads running (e.g. device driver threads, GUI threads, background tasks, etc.) Even when the system is idling (0% CPU load), these threads will get the CPU every now and then, and cause outliers.

7.3.6. Outlier detection

Fortunately, outliers can be reliably detected. First, the raw sequence $t(0), \ldots, t(N-1)$ is trimmed by the prerun values and then the arithmetic mean value $\mu = N^{-1} \sum t(i)$ and standard deviation $\sigma = [N^{-1} \sum (t(i) - \mu)^2]^{1/2}$ are computed. For a predefined ratio S, all values $t(i)$ that are below or equal the threshold $t(i) \leq \mu + S\sigma$ are selected as valid samples, forming a sequence $t'(0), t'(1), \ldots, t'(M-1)$. Values $t(i) > \mu + S\sigma$ above this limit are discarded as outliers. A too little threshold S removes tolerable samples from the sequence. The threshold S may not be too large either, as for a noisy sequence too many faulty samples are regarded as valid. Empirical examinations showed, that a ratio $S = 3$ is a good compromise for a large variety of runtimes—ranging from short to long operations. In all examined cases, the threshold selected $> 99\%$ of the samples as valid. Figure 7.1 shows a histogram of the distributions of sample values for the measurement of a 65536-point real-data FFT. Obviously, measurement samples in the benchmark process are not normally distributed. Regarding the mean value μ, the lower interval $[0, \mu)$ always contained less samples than the upper interval $[\mu, \mu + S\sigma]$. The following interpretation can be used: For each operation and size there must exist a deterministic lower bound for the runtime, which relates to the number of instructions of the code, the processor and memory architecture, etc. Generally speaking, the actual runtime on the system results can be understood as the minimum runtime plus additional disturbances (noise), caused by the effects discussed above. The minimum runtime is measured infrequently, as regularly the execution is non-optimal and subject to delays. Very strong disturbances (outliers) are very unlikely either. Typical stochastic properties of measurement sequences are shown for three examples in table 7.1. Eventually, each measurement has to be represented by a single value. Therefore the arithmetic mean value $\mu^* = M^{-1} \sum t'(i)$ of the outlier-filtered sequence is calculated and stored in the performance profile. These mean values are thus later used in the evaluation of the cost functions.

7.3.7. Robustness

Considering that a benchmark of the same operation of identical size is executed several times over, the results should be very similar without significant variations. It should be regardless, if the benchmark is executed several minutes, hours or days later. The performance measures must be reproducible and the benchmark process shall be *robust*. Therefore, a prerequisite is that the target system is in a comparable state for all times the benchmark is performed. It may not be idling for one run, where at another point it has some background load. However, when these preconditions are met, any larger variances indicate a faulty benchmark procedure.

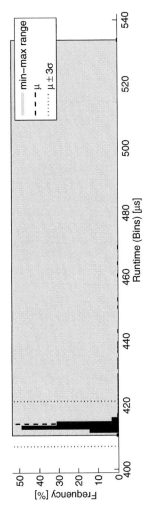

Figure 7.1.: Measurement statistic for a 65536-point real-data FFT

Size	min [μs]	max [μs]	mean μ [μs]	std σ [μs]	med [μs]	outlier-filtered mean μ^* [μs]	valid [%]	outliers [%]
128	0.6024	1.5303	0.6105	0.013	0.6061	0.6103	100.00	0.000
1024	3.3411	9.4247	3.3795	0.074	3.3897	3.3783	99.970	0.030
65536	410.69	533.72	414.24	2.342	414.11	414.11	99.289	0.716

Table 7.1.: Stochastic properties of three real-data FFT measurements

205

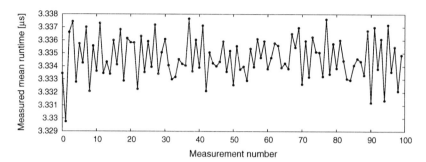

Figure 7.2.: Mean runtimes of 100 separately executed
benchmarks of a 1024-point FFT

The presented benchmark method has been verified for robustness for several operations and sizes. While ensuring that the system was idle, the same operation and size has been benchmarked a 100 times over. The resulting deviations were surprisingly small. Figure 7.2 illustrates this with the example of the 1024-point real-data FFT.This sequence has a min-max span $\Delta = 7.9\,\text{ns}$, a mean value $\mu = 3.335\,\mu\text{s}$ and standard deviation $\sigma = 1.59\,\text{ns}$. This corresponds to a relative standard deviation $\sigma/\mu \approx 0,24\%$. Similar results were found for other sizes, without any example of significant deviations. This proved that the benchmark method is sufficiently robust.

7.4. Basic operations

All fast convolution algorithms involve fundamental arithmetic operations, like additions, subtraction, multiplication and divisions. Rarely are these binary operations are executed for two operands (*scalars*) only. In the majority of cases, the operands are vectors of the same number of elements (e.g. blocks of samples or spectra). On modern processors these operations can be *vectorized* (instruction-level parallelism), increasing the arithmetic throughput. discrete Fourier transform (DFT)-based convolution methods demand complex-valued arithmetic. Present general purpose processors do not support complex-valued arithmetic natively. Hence, it has to be realized by sequences of elementary real-valued arithmetic operations. In recent years, the SIMD support (e.g. SSE1-SSE5, AVX, AVX2, etc.) advanced over time. Particularly, complex-valued multiplications benefit from these enhancements. Today, the theoretic number of arithmetic operations is only an indicator for complexity of an operation. It typically fails to precisely predict the actual runtimes on a target machine. Single arithmetic operations can not always translate into single instructions anymore. Nowadays, powerful software libraries (e.g. Intel Performance Primitives) provide convenient high-level arithmetic operations in software and programmers do not have to hassle with the implementation (e.g. in assembly language) themselves. Often it is beneficial to use combined instructions, which realize two operators at once. This increases the arithmetic density of the code and can save memory access instructions (load/store). In practice, these combined operations often turn out to be faster than two individual operations. An operation of particular importance for partitioned FFT-based convolution algorithms is *complex-valued multiply-accumulate* (CMAC). It performs a complex-valued multiplication and adds the result to a destination variable. The following arithmetic operations are considered in this work.

- **ADD**: real-valued in-place vector addition
  $x(i) = x(i) + y(i) \quad (x(i), y(i) \in \mathbb{R})$

- **MUL**: real-valued in-place point-wise vector multiply
  $x(i) = x(i) \cdot y(i) \quad (x(i), y(i) \in \mathbb{R})$

- **CMUL**: complex-valued in-place point-wise vector multiply
  $x(i) = x(i) \cdot y(i) \quad (x(i), y(i) \in \mathbb{C})$

- **CMAC**: complex-valued in-place point-wise vector multiply-accumulate
  $x(i) = x(i) + y(i) \cdot z(i) \quad (x(i), y(i), z(i) \in \mathbb{C})$

In-place operations are considered here. For reasonably large data, they are often slightly faster than *out-of-place* operations, where the result is stored in a different variable. All operations listed above were benchmarked on the test system. They were realized using Intel's Performance Primitives library version 7.0 as vectorized in-place operations on 32-bit floating points. Unless stated otherwise, real-valued operands are considered.

Figure 7.3 gives an overview about the computational complexity of the operations on the target system. For an easier comparison, the diagram depicts the average number of CPU cycles *per point* (divided by the transform size). On the logarithmic-scale abscissa, the typical $\mathcal{O}(N \log N)$ runtime of the FFT shows up as a line, whereas the other $\mathcal{O}(N)$ operations are represented by a constant value. The operations with the lowest costs are real-valued arithmetic operations. Complex-valued operations are significantly more expensive. For sizes $> 5k$ the complex-valued multiplications were $\approx 2.4\times$ more expensive than their real-valued counterparts. Function call and loop overheads penalize computation on small operands and cause a slight increase in the costs per point. Yet in respect to fast Fourier transforms, the arithmetic operations are comparably cheap. Here, the costs per point of short FFTs of length $N \leq 512$ are not strictly increasing, which contradicts the theoretic behavior. The advantage of combined operations become directly visible for the CMAC operation (which involved two real additions), which is even cheaper than the sum of CMUL and ADD (the latter including only one real addition). The combined complex-valued multiply-add (CMAC) is $1.2 - 1.6\times$ faster than two individual operations. For larger sizes $N > 4000$ the speedup converged to $\approx 1.5\times$. For real-valued operations cache losses begin affecting the performance for $N \approx 4k$ (16kB per operand) and increase the computation time by a factor of ≈ 1.7. For complex-valued operations this 'step' was identified at $N \approx 1.5k - 2k$, resulting in $\approx 1.3\times$ longer execution times.

Discrete transforms

The runtimes of a variety of discrete transforms has been benchmarked on the test system, including

- Fast Fourier Transforms of complex data (**FFT-C2C, IFFT-C2C**)

- Fast Fourier Transforms of real data (**FFT-R2C, IFFT-C2R**)

- Fast Hartley Transforms (**FHT**)

- Fast Discrete Cosine Transforms (**DCT1, DCT2, DCT3, DCT4**)

- Fast Discrete Sine Transforms (**DST1, DST2, DST3, DST4**)

All of them are in-place transforms of 32-bit floating points. They were computed using FFTW 3.3.3 and planned using the flag *FFTW_MEASURE*. The results for power-of-two transform sizes are shown in figure 7.4. Computational costs are measured by the average number of CPU cycles divided by the transform size. Obviously, there are large differences in execution speeds between the different transforms. The transforms can be classified by execution speed in ascending order:

1. Real-data FFTs/IFFTs (FFT-R2C, IFFT-C2R) compute fastest

2. Complex-data FFTs/IFFTs (FFT-C2C, IFFT-C2C) are between $\approx 1.25 - 1.85\times$ slower

3. Fast Hartley transforms (FHTs) are $> 2\times$ slower than real-data FFTs for sizes $\geq 1k$. For shorter sizes < 256 they can outperform complex-data FFTs

4. Fast discrete cosine/sine transforms are even slower than FHTs (observed order: DCT/DST-2 $<$ DCT/DST-3 $<$ DCT/DST-4 $<$ DCT/DST-1. DCT/DST-1 transforms computed slowest[5] and fall out of range in figure 7.4)

---

[5]See explaination in the FFTW3 manual

In-place and out-of-place transforms had comparable execution times. For small transform sizes, in-place transforms compute usually a bit faster. Out-of-place transforms can be slightly faster for large sizes. There can be marginal speed differences between a forward FFT and an inverse FFT.

7.4.1. Cost relations

The optimization of convolution algorithm is strongly affected by the relations of the costs of different operations. The cost ratio between FFTs and spectral convolutions is thereby of particular importance. If the relative costs between different operations are similar in theory and practice, it is probable that theoretically optimal choices of parameters may perform well in practice. In order to compare these ratios, the costs of operations were described using simple runtime models. Linear-time $\mathcal{O}(()N)$ operations using a runtime function of the form $T(N) = kN$ and fast $\mathcal{O}(N \log N)$ transforms by a term $T(N) = kN \log_2 N$. The constants k are regarded in the following.

All basic operations ADD, MUL, CMUL and CMAC have a linear runtime in $\mathcal{O}(N)$, where N is the size of the vector operands (number of elements). In theory they require the following number of (real-valued) arithmetic operations

$$T_{\mathrm{ADD}}(N) = T_{\mathrm{MUL}}(N) = N \tag{7.1}$$

$$T_{\mathrm{CMUL}}(N) = 6N \tag{7.2}$$

$$T_{\mathrm{CMAC}}(N) = T_{\mathrm{CMUL}}(N) + T_{\mathrm{CADD}}(N) = 8N \tag{7.3}$$

Two complex numbers are added by separate addition of their real and imaginary parts, which requires two real additions. The product of two complex numbers can be realized using six arithmetic operations, four real-valued multiplications and a real-valued subtraction and a real-valued addition

$$(a + bi) \cdot (c + di) = \underbrace{[ac - bd]}_{\text{2 muls,1 sub}} + \underbrace{[ad + bc]}_{\text{2 muls,1 add}} i \tag{7.4}$$

An alternative form saves one multiplication at the cost of three further additions/subtractions, resulting in seven operations

$$(a + bi) \cdot (c + di) = \underbrace{[\underbrace{ac}_{=T_1} - \underbrace{bd}_{=T_2}]}_{\text{2 muls, 1 sub}} + \underbrace{[(a + b)(c + d) - T_1 - T_2]}_{\text{2 adds, 1 mul, 2 subs}} i \tag{7.5}$$

As the real operations require the same number of CPU cycles on most processors, this form is not of an advantage anymore. Each operation FFTC2C, IFFTC2C, FFTR2C and IFFTC2R has different constants k. Here, the con-

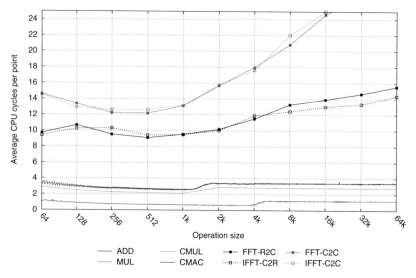

Figure 7.3.: Computational complexity of basic operations on the test system (arithmetic operations computed using Intel IPP and power-of-two transforms using FFTW)

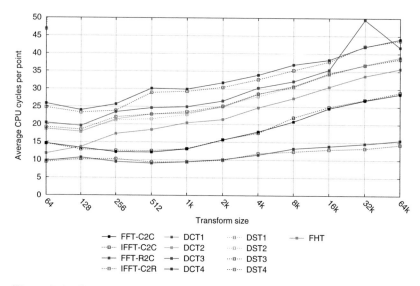

Figure 7.4.: Computational complexity of discrete transforms computed (power-of-two sizes) on the test system (computed using FFTW)

¹Implemented using Intel Performance Primitives
²Power-of-two real-data split-radix FFT by Johnson and Frigo [50]
³Power-of-two complex-data split-radix FFT by Johnson and Frigo [50]

Operation	Cost model	Parameters		
		Theory	Practice[1]	Range
ADD	$T_{\text{ADD}}(N) = k \cdot N$	k=1	$k \approx 0.69$ $k \approx 1.17$	$(128 \leq N \leq 1k)$ $(2k \leq N \leq 64k)$
MUL	$T_{\text{MUL}}(N) = k \cdot N$	k=1	$k \approx 0.69$ $k \approx 1.17$	$(128 \leq N \leq 1k)$ $(2k \leq N \leq 64k)$
CMUL	$T_{\text{CMUL}}(N) = k \cdot N$	k=6	$k \approx 2.16$ $k \approx 2.82$	$(128 \leq N \leq 1k)$ $(2k \leq N \leq 64k)$
CMAC	$T_{\text{CMAC}}(N) = k \cdot N$	k=8	$k \approx 2.61$ $k \approx 3.40$	$(128 \leq N \leq 1k)$ $(2k \leq N \leq 64k)$
FFT-R2C	$T_{\text{FFT-R2C}}(N) = k \cdot N \log N$	$k \approx 1.68^2$	$k \approx 0.97$	$(16 \leq N \leq 64k)$
FFT-C2C	$T_{\text{FFT-C2C}}(N) = k \cdot N \log N$	$k \approx 3.49^3$	$k \approx 1.79$	$(16 \leq N \leq 64k)$

Table 7.2.: Cost models and parameters in theory and on the test system in practice

stants were obtained by curve-fitting the exact number of arithmetic operations of the split-radix FFTs by Johnson and Frigo [50]. For simplicity it is assumed that an inverse FFT has the same costs as a forward FFT. In order to obtain constants for the practical case, curve-fitting was as well performed on the performance profile of the test system. Table 7.2 lists the runtime constants k for all regarded operations, in theory and practice. For a better distinction, two separate values are provided for the arithmetic operations, with respect to the cache limits (see section 7.4). They illustrate the influence of the effective caching in simple numbers. Complex and real multiplies have in theory a relation of $k_{\mathrm{CMUL}}/k_{\mathrm{MUL}} = 6/1 = 6$. In practice however, the ratio is approximately $2.82/1.17 \approx 2.41$. Here, complex-valued arithmetic is in relation significantly cheaper. The ratio between real-data FFTs and complex-valued multiplications is about $k_{\mathrm{FFTR2C}}/k_{\mathrm{CMUL}} \approx 1.68/6 = 0.28$ in theory. In practice however, the ratio is approximately $0.97/2.82 = 0.344$. The observed differences show, that the theoretical cost relations of the operations do not necessarily apply on the target system.

7.5. Accuracy

The accuracy of the proposed semi-empirical benchmark-based cost model was verified with a representative algorithm. The algorithm should involve a meaningful number of elementary operations and not consist of a few function calls only. Simple OLA and OLS methods without filter partitioning (see chapter 4) are hence rather unsuited for verifying the approach. Instead, uniformly-partitioned FFT-based Overlap-Save (OLS) convolution (UPOLS) was selected. This algorithm, introduced in section 5, involves a larger number of other basic operations, in particular CMAC. For short filters it converges into the unpartitioned convolution algorithm and covers this case as well. Moreover, the algorithm is the core of non-uniformly partitioned convolution techniques, which are used for efficient filtering with long impulse responses.

A reference implementation of uniformly partitioned Overlap-Save (UPOLS) was created and then benchmarked on the test system. The implementation uses a standard transform size $K = 2B$ of twice the block length and real-data transforms. Using the previously introduced benchmark technique, the average computation time for the stream processing of a single length-B block was recorded. Afterwards, this measure was converted into CPU cycles and divided by the block length N. This performance values were compared to the prediction using the semi-empirical cost model.

The computational complexities per output sample for the measurement and prediction are plotted in figure 7.5. The relative prediction error $(T_{\mathrm{meas.}} - T_{\mathrm{pred.}})/T_{\mathrm{meas.}}$ is shown in figure 7.6. Clearly, the accuracy of the estimation depends on the data sizes of the operations and in particular the block length B. Convolutions that involve small-sized elementary operations ($B \leq$ 1k) are subject to larger prediction errors. Here, the prediction always underestimates the costs. The relative deviations reach up to 17% towards longer filters. It can be observed, that for short filter lengths N they are significantly smaller. This is important to mention, because sub filters in non-uniformly partitioned convolution algorithms have larger block lengths B and comparably small lengths N. Thus, the prediction error is effectively smaller for these algorithms. The assembly of these very small operations in the convolution algorithm is stronger affected by low-level cache misses and memory accesses. A close examination revealed, that even the combined runtimes of a small FFT and inverse fast Fourier transform (IFFT) were a little larger than the sum of runtimes of the independently measured operations. Also is the complex-valued multiply-accumulate (CMAC) operation slightly slower, when integrated into the algorithm. This affects the slopes of the graphs and the offset in the relative errors. Using a linear correction term kN with a constant k the model could be matched to the measured data quite well, resulting in lower errors. This indicates, that the errors stem from the size of operands and their memory requirements. When elementary operations of larger sizes ($B \geq$ 1k) are involved, longer computation times occur and the prediction is very accurate. The total costs per sample are nevertheless low, because they are divided by larger block lengths B.

It can be summarized, that the cost model reproduces the propagation of the computational complexity quite well. Moderate prediction errors to the actual runtimes occur in the case of small elementary operations (short block length B and short filters N). The proposed cost model is conceptually simple. It does not regard cache utilization and memory accesses. As seen above, the influence of these factors can be relatively large, when the involved operations are tiny ($<$ 1000 CPU cycles). In this respect the overall results are good. The model could be further improved by considering these factors as well. However, the modeling of these influence factors is not trivial.

7.6. Efficient FFTs

The computation time of a fast Fourier transform strongly depends on the transform size. Trivially, a large FFT computes slower than a shorter one. But the operations per point (runtime divided by transform size) might be less for the larger transform. Mainly the prime factors of the transform size determine which decompositions are applicable within the FFT computation (e.g. CFM or PFM, for details see section 2.5.2). This has a significant impact on the runtime of an FFT. For instance, a 256-point (power-of-two) real-data FFT has a runtime of 1.11 μs on the test system. For a transform

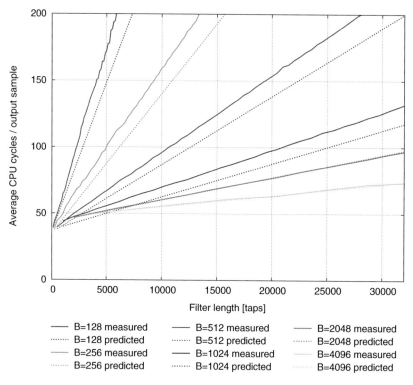

Figure 7.5.: Measured and predicted algorithmic complexity of UPOLS stream processing for different block lengths on the test system

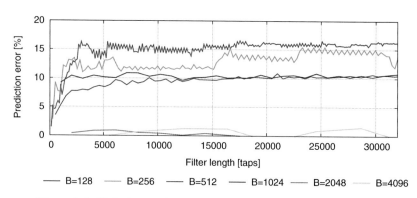

Figure 7.6.: Relative prediction error of UPOLS stream processing for different block lengths on the test system

size of 257 (which is a prime) the same operation consumed 21.75 μs —which is more than 20× slower. Fast $\mathcal{O}(N \log N)$ FFT algorithms are available, also for prime N. Precise mathematical formulas for the number of operations of an FFT can be bulky (for example in [50]), which make it cumbersome to use them for a theoretical analysis. Often the following simplified model is favored for expressing the computational costs of an FFT

$$T_{\text{FFT}}(N) = k \cdot N \log_2 N \qquad (k \in \mathbb{R} \text{ const.}) \qquad (7.6)$$

Here, the scaling factor k depends on the transform type. Note, that the form in 7.6 neglects additive terms and can therefore not be *exact*. Restricted to power-of-two sizes, it is however a good approximation with comparably little errors for smaller N. Table 7.2 provides several values for the proportionality constant k in theory and practice. The constant k can also serve as a measure for the efficiency, by rearranging Eq. 7.6 and making it a function $k(N)$ over N

$$k(N) = \frac{T_{\text{FFT}}(N)}{N \log_2 N} \qquad (7.7)$$

Eq. 7.7 puts the actual number of cycles in relation to theoretical runtime complexity[4]. Small values of k indicate relatively low runtimes (high efficiency) and vice versa. The most and least efficient transforms can be determined by evaluating Eq. 7.7 for all measured runtimes $T_{\text{FFT}}(N)$ and by selecting the smallest respectively largest values.

The runtimes of all common FFT types have been extensively benchmarked on the test system. This included *all* sizes from 1 to 65536. Figure 7.7 shows the results of the real-data FFTs. The span in runtimes is wide. Two cost functions (Eq. 7.6) with factors $k = 1$ and $k = 40$ have been added as guidelines. Transforms of highly composite sizes, like $640 = 2^7 \cdot 5$, compute very efficient. In contrast, prime-size FFTs have the longest runtimes. Between these cases, the scaling factors can differ by a factor of 40. This proves evidence, that the computational complexity of an FFT can in general not be expressed by Eq. 7.6 and the mathematical properties of the transform size must be considered.

Table 7.3 list the top twenty most and least efficient transforms in the measurements. Note, that this list contains the best and worst candidates in the entire range of sizes within $1 \leq N \leq 65,536$. For a better overview, powers of two are marked gray. The findings support the claim stated above. The sizes of the most efficient real-data FFTs (left columns) contain powers-of-two as a factor. Only the small primes 3 and 5 are found as additional factors, if N is not a power-of-two. The most inefficient transforms are those of prime sizes (right columns) and they should be avoided.

---

[4]The authors of FFTW use the reciprocal of $k(N)$ as a performance measure [36]

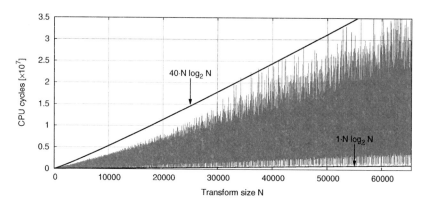

Figure 7.7.: Measured number of CPU cycles of real-data FFTs for all sizes $1 \leq N \leq 64\mathrm{k}$

However, it can also be seen, that the list of most efficient transforms is not assembled from power-of-two sizes exclusively. Inspecting table 7.4) provides an answer. It shows the efficiency for power-of-two sizes only. A general observation is, that for smaller transform sizes $N \rightarrow 1$ the factor $k(N)$ increases and thus, the relative efficiency drops. The reason is presumed the fixed function call overhead, which an relation becomes large for smaller transforms with few instructions. For power-of-two transform size $N \geq 512$, the simplified cost function (Eq. 7.6) is a good approximation, as $k(N)$ varies only little.

A more detailed look on the performance is gained by counteracting the general trend of efficiency decrease towards smaller transform sizes N. This is achieved by inspecting sub ranges of the transform size and not the full range at once, as in table 7.3. Table 7.5 list the three most efficient transforms within such specific sub ranges. Here it can be seen, that indeed powers-of-two are the first or second most efficient transforms in their groups.

Which transform size is actually worth the consideration can only be answered in the context of an application. The preceding chapters showed, that several convolution algorithms can strongly benefit from using untypical transform sizes, which are not powers of two.

Most efficient			Least efficient		
Size	$k(n)$	Prime factors	Size	$k(n)$	Prime factors
1,280	0.910	$(2^8\, 5^1)$	37,549	38.629	$(37{,}549^1)$
20,480	0.916	$(2^{12}\, 5^1)$	33,377	38.652	$(33{,}377^1)$
2,048	0.925	(2^{11})	38,053	38.784	$(38{,}053^1)$
768	0.927	$(2^8\, 3^1)$	40,193	38.800	$(40{,}193^1)$
2,560	0.930	$(2^9\, 5^1)$	43,961	39.169	$(43{,}961^1)$
1,024	0.953	(2^{10})	42,391	39.297	$(42{,}391^1)$
25,600	0.953	$(2^{10}\, 5^2)$	39,251	39.387	$(39{,}251^1)$
4,096	0.958	(2^{12})	52,813	39.417	$(52{,}813^1)$
1,920	0.960	$(2^7\, 3^1\, 5^1)$	53,441	39.446	$(53{,}441^1)$
3,072	0.960	$(2^{10}\, 3^1)$	41,729	39.621	$(41{,}729^1)$
49,152	0.961	$(2^{14}\, 3^1)$	56,113	39.795	$(56{,}113^1)$
64,000	0.961	$(2^9\, 5^3)$	50,101	39.992	$(50{,}101^1)$
3,200	0.962	$(2^7\, 5^2)$	45,641	40.019	$(45{,}641^1)$
15,360	0.963	$(2^{10}\, 3^1\, 5^1)$	32,971	40.282	$(32{,}971^1)$
1,536	0.964	$(2^9\, 3^1)$	41,077	40.436	$(41{,}077^1)$
2,400	0.964	$(2^5\, 3^1\, 5^2)$	37,409	40.981	$(37{,}409^1)$
10,240	0.969	$(2^{11}\, 5^1)$	36,073	41.030	$(36{,}073^1)$
65,536	0.972	(2^{16})	34,231	41.161	$(34{,}231^1)$
46,080	0.973	$(2^{10}\, 3^2\, 5^1)$	63,127	41.337	$(63{,}127^1)$
32,768	0.977	(2^{15})	58,451	43.302	$(58{,}451^1)$

Table 7.3.: Real-data FFTs with the highest and the lowest computational efficiency for sizes the interval $1 \leq N \leq 65.536$ (power-of-two sizes are marked gray)

Size	Efficiency	Runtime [μs]	Size	Efficiency	Runtime [μs]
8	5.524	0.055	1,024	0.953	4.057
16	2.981	0.079	2,048	0.925	8.659
32	1.986	0.132	4,096	0.958	19.58
64	1.615	0.258	8,192	1.021	45.20
128	1.524	0.568	16,384	0.995	94.86
256	1.189	1.012	32,768	0.977	199.6
512	1.010	1.935	65,536	0.972	423.6

Table 7.4.: Computational efficiency of power-of-two real-data FFTs

Size interval	Size	$k(n)$	Prime factors	Size interval	Size	$k(n)$	Prime factors
$1 < N \leq 16$	16	2.981	(2^4)	$128 \leq N \leq 256$	256	1.189	(2^8)
	15	3.547	$(3^1,5^1)$		240	1.374	$(2^4,3^1,5^1)$
	12	3.753	$(2^2,3^1)$		200	1.465	$(2^3,5^2)$
$16 < N \leq 32$	32	1.986	(2^5)	$256 < N \leq 512$	512	1.010	(2^9)
	20	2.622	$(2^2,5^1)$		384	1.103	$(2^7,3^1)$
	30	3.152	$(2^1,3^1,5^1)$		448	1.132	$(2^6,7^1)$
$32 < N \leq 64$	64	1.615	(2^6)	$512 < N \leq 1,024$	768	0.927	$(2^8,3^1)$
	40	2.542	$(2^3,5^1)$		1,024	0.953	(2^{10})
	50	2.798	$(2^1,5^2)$		960	1.029	$(2^6,3^1,5^1)$
$64 < N \leq 128$	128	1.524	(2^7)	$1,024 < N \leq 2,048$	1,280	0.910	$(2^8,5^1)$
	120	1.744	$(2^3,3^1,5^1)$		2,048	0.925	(2^{11})
	96	1.847	$(2^5,3^1)$		1,920	0.960	$(2^7,3^1,5^1)$

Table 7.5.: Top three most efficient real-data FFTs within specific intervals of sizes (power-of-two sizes are marked gray)

7.7. Conclusions

The presented benchmark-based cost models assess the computational complexity of algorithms with respect to an actual target system. This proved necessary, as the relative costs of operations in practice differ from those in theory. It is therefore advised to optimize the convolution algorithm based on this data. A theoretically optimal choice of parameters may not be the best choice in practice. The benchmark-based cost model reproduces the progression of costs quite well. However, its overall accuracy depends on the number and size of involved operations. A limitation of the model is its simplicity, neglecting aspects like for instance memory accesses and cache-utilization. Particularly, when low-complexity operations are used, which only consume a little number of cycles on the processor, side-effects like call overheads, loop instructions and probably also the cache utilization of low-level caches affect the quality of the cost prediction. In these cases, the sum of runtimes of the individual operation underestimates the actual runtime on the machine. As soon as the operations get more complex, the prediction becomes very accurate. However, the overall quality of the model is considered sufficient for the study and analysis of convolution algorithms composed from the enumerated elementary operations. The accuracy of the cost prediction can be further improved by enhancing the model or by partly substituting the predicted costs with actual measurements. Whether this is possible depends on the application and the number of parameter combinations.

8. Conclusion

This thesis' objective was the research of finite impulse response (FIR) filtering technologies for applications in acoustic virtual reality. The considerations aimed for standard processors and a block-wise audio processing exclusively. The diverse requirements of the field (e.g. multiple channels, time-varying filters) demand special algorithms. Conventional fast convolution techniques must be adapted to serve for a real-time processing, otherwise the processing delay might be unacceptably large. Among different divide-and-conquer approaches, partitioned convolution was identified as the strategy which accords real-time processing. It decomposes the operands into interconnected sub sequences (blocks) along the time axis. This makes it possible to compute parts of the filtering in accordance with the availability of input samples. The latency can be kept low, which allows realizing the short response times needed in virtual reality. Partitioned convolution algorithms are mainly influenced by two factors: the underlying fast convolution technique and the partitioning of the operands. Both factors have been carefully examined in this thesis.

8.1. Summary

The examination of the wide class of conventional convolution algorithms in chapter 4 confirmed the status of FFT-based convolution as a reference method. Realizing a fast convolution in the DFT domain is conceptually simple. Benchmarks showed that current high-performance FFT algorithms outperform potential alternative transforms. For short filters, interpolation-based convolution techniques, like the Karatsuba algorithm, might turn out as an alternative approach—benchmarks of matured implementations are needed for further clarification. Similarly, the performance of number theoretic approaches implemented on general purpose processors is unverified. At least in theory these methods could match up with the FFT. However, it has also been shown, these algorithms can not outperform FFT-based convolution for long filters. The conjunction of transform-based fast convolution and partitioned convolution is particularly attractive. Huge computational savings are achieved by implementing most of the processing in the frequency-domain and using transforms economically.

8. Conclusion

Partitioned convolution techniques are manifold and characterized by different properties. Conventional fast convolution techniques are generally made applicable for real-time filtering using the Overlap-Add (OLA) or Overlap-Save (OLS) scheme. Technically, this already marks a partitioned convolution, as the input signal is processed in blocks of equal sizes. In the examinations, an FFT-based OLS convolution allows speedups over a time-domain FIR filter in the range of $5 - 80\times$, depending on the block length used. Especially for regular OLA/OLS convolution (without filter partitioning) it is advised to consider uncommon transform sizes, e.g. small multiples of powers-of-two. Restricting the choice exclusively on powers of two can significantly increase the computational costs. A time-domain processing should only be considered in the case of very short filters of less than 32 taps. Here, short linear convolution algorithms (e.g. the Karatsuba algorithm) might still be faster.

Besides the already partitioned signal, the second operand, the filter, can be partitioned as well. The partitioning of the filter marks a free parameter, unlike the partitioning of the signal, whose block length is determined by the latency requirements. Elementary techniques, which do not partition the filters, suffer a common drawback: They become relatively inefficient for the case that one operand is significantly longer than the other. Such unbalanced convolutions often occur in acoustic virtual reality and particularly when reverberation is realized using convolution. Partitioning can be interpreted as a divide-and-conquer approach. By splitting the filter into uniform sub filters, these on their own are more balanced and can be more efficiently processed. In total, this allow a significant reduction of the computational effort.

An essential improvement arises transform-based fast convolution is combined with a uniform filter partitioning: Given, that signal and filter are partitioned with the same granularity (the common choice), the sub filters can be realized entirely in the frequency-domain. Each processed input and output block needs to be transformed only once. The involved fast Fourier transforms become independent of the filter length N, making the class of methods asymptotically faster than the elementary techniques. For a 4000-tap filter and block lengths in the range of 128-1024 samples, speedups between $2\times -8\times$ were identified on the test system. It is remarked, that the uniform partitioning of the filter also moderately increases the computational effort for transforming the impulse response.

The presented generalization of uniformly partitioned convolution supported independent partitions in both operands to be realized with arbitrary transform sizes. These further degrees of freedom were introduced for evaluating the complexity bounds of uniformly partitioned convolution. It showed, that the standard parameters (equal partitioning in both operands) indeed mark the optimal solution in many cases. As a rule of thumb, this holds for filter lengths in the approximate range $B \leq N \leq 20 \cdot B$. Given, that the block length is a power of two (the usual case), power-of-two transforms mark the best choice.

The computational costs for filtering with partitioned convolution techniques depend on the block length of the audio processing and hence on the input-to-output latency. For the classes of unpartitioned and uniformly-partitioned filters, the relation is nearly anti-proportional, unless the filters are short. This can be summarized in a simple statement: Halving the block length (which does not necessarily mean that the latency is halved) doubles the computational effort of the filtering.

The motivation for a non-uniform partitioning of the filter impulse response originates from the observation, that later sub filters can be implemented with longer block lengths. This once more allows a significant reduction of the computational costs. Moreover, the steady increase of block lengths results in a lower runtime complexity class. For the example case of two seconds of reverberation time, the non-uniform techniques outperformed the uniform approch by a factor of 10. For longer filters and short block lengths, the savings get even larger.

However, these computational advances come at a price: Non-uniformly partitioned convolution required much more effort in its implementation. All previously discussed methods feature a recurrent execution pattern, with similar amounts of computation. In contrast, achieving an even load distribution with non-uniformly partitioned convolution is not trivial. Also is the filter exchange limited, which is of particular relevance in real-time auralization. All these aspects are influenced by the partitioning, which marks a key parameter of the algorithms. The strong point of Gardner's partitioning scheme is its practicality. However, it does not result in minimal computational costs. Specifically optimized partitions fully exploit the specific properties of uniformly-partitioned convolutions (i.e. frequency-domain delay-lines) and can almost half the effort over Gardner's approach. A sheer optimization for minimal algorithmic costs results in impractical partitions with a worst-case load distribution. It was presented how the optimization can be guided to obey a minimal flexibility for the scheduling of sub convolutions. The increase in computational costs over the theoretical minimum is moderate and still a huge leap forward. Considerations on the design of non-uniform partitions for real-time room acoustics were conducted. A good compromise between the requirements (computational effort, real-time scheduling, time-varying filters) can be achieved.

8.2. Guidelines

The findings of this thesis lead to the following guidelines for real-time filtering in acoustic virtual reality. They are classified by aspects.

Performance

- Time-domain FIR filtering is only advised for very short filters. The theoretical break-even point where FFT-based convolution becomes faster is between 16-32 samples (cp. Tab. 2.1). In practice, this boundary depends on the implementations.

- Short linear convolution algorithms (e.g. Karatsuba algorithm) outperform the naive time-domain approach. At least in theory, they are an alternative to FFT-based convolution for filter length $N \leq 128$ (cp. Tab. 2.1). However, the performance of the Karatsuba technique is unverified.

- The computational effort of all following algorithms depends on the block length B (latency). Break-even points in filter lengths N are expressed as multiples of B.

- Conventional Overlap-Add (OLA) or Overlap-Save (OLS) FFT-based convolution is suggested for the case of very short filters $N \leq B$. Here, both schemes are similar in their performance. The simplicity of the Overlap-Add (OLA)/Overlap-Save (OLS) algorithms is an argument for using them for longer filters also. In this case non power-of-two transform sizes should be considered. However, Overlap-Add (OLA)/Overlap-Save (OLS) convolution without filter partitioning is significantly less efficient than uniformly-partitioned techniques.

- Uniformly-partitioned convolution is the method of choice for short to medium length FIR filters in the range $B < N < 20 \cdot B$. The standard parameters (filter partition $L =$ block length B, transform size $K = 2B$) deliver an optimal or near-to-optimal performance for common power-of-two block lengths B. Only given filters longer than $N > 20 \cdot B$, the use of individual optimized transform sizes and separate partitions in signal and filter is beneficial. However, non-uniformly partitioned techniques should be considered for these filter lengths, as they will possibly deliver a better performance.

- For filters exceeding $N > 20 \cdot B \dots N > 20 \cdot B$ a non-uniform filter partitioning should be considered. The computational savings over uniformly partitioned techniques grows over the filter length. This accords in particular for low-latency processing, as they increase strongly for towards short block length B. For artificial reverberation (e.g. filters of 500 ms . . . 2 s reverberation time) the savings can reach several magnitudes (cp. Fig. 5.6).

Benchmark-based cost models proved sufficient to quantitatively assess the algorithms' computational costs. For precise, qualitative statements the actual algorithm should be benchmarked on the target machine.

Implementation

- Regular OLA and OLS convolutions are very simple technique which can be implemented in a few lines of code. fast Fourier transforms dominate the computation. Partial transforms deliver too little savings compared to their implementational burden.

- Standard uniformly-partitioned convolution methods (UPOLA or UPOLS) are conceptually simple algorithms, demanding a little more effort in their implementation, which is greatly rewarded by the computational savings. The techniques compute short transforms only, which consume a minor part of the computations only. Most of the computations are complex-valued multiplications and accumulations, which strongly affect the algorithms' performance.

- Non-uniformly partitioned convolution techniques requires significantly more implementation effort and programmer skills. In principle, non-uniformly partitioned Overlap-Add (NUPOLA)/non-uniformly partitioned Overlap-Save (NUPOLS) convolution engines can be assembled from uniformly partitioned Overlap-Add (UPOLA) or UPOLS units. The difficulties of an implementation are the time-critical scheduling of sub convolutions—in particular on multiprocessor systems. Gardner's partitioning scheme [39] is inheres relatively loose timing constraints and results in a periodic scheduling. A time-invariant filtering can be realized using a hard-wired, static execution plan for the sub convolutions. If the filtering is subject to changes (e.g. modification of channels or filters) a dynamic scheduling is strongly suggested. For the best performance—not only considering the load—the non-uniform filter partition should be optimized using benchmarks. A hand-tuned partitioning might be more economically considering the effort of this procedure. Non-uniformly partitioned convolution should considered with care, if an even distribution of the computational load is essential or when a multi-threaded programming is impossible. The computation in these techniques is shared by FFTs and spectrum multiplications and additions, making both important.

Latency

Apart from time-domain FIR filters, all regarded algorithms revealed a dependency between their computational effort and the input-to-output latency. For unpartitioned and uniformly-partitioned techniques an almost anti-proportional relation was observed. Only non-uniformly approaches behave differently and show a logarithmic dependency. In other words, a lower latency is achievable with these techniques at a relatively low cost.

The filter exchange latency is affected by the computation time for the filter transformation. For most applications this runtime will be small compared to other parts of the software. The partitioning of the filter increases the costs of the filter transformation. However, it also opens the possibitliy for parallelization.

The OLA/OLS respectively UPOLA/UPOLS methods allow frequent filter updates with the frame rate of the audio stream. In non-uniformly partitioned methods the filter exchange is affected by the filter partition. It can limit the maximum update rates and introduce additional latency. (see section 6.11).

Filter networks

Networks of FIR filters, including filters with multiple inputs and outputs, can be significantly accelerated. Such networks occur often in auralization systems—e.g. multi-channel binaural synthesis, crosstalk cancellation filters or room acoustic audio rendering. Large computational savings can be achieved by implementing most of the processing in the frequency-domain, saving unnecessary transforms. Parallel structures are easy and straightforward to implement. Cascades of multiple FIR demand care: The transform size must be chosen large enough to prevent time-aliasing in the assembly. For sequences of non-uniformly partitioned convolutions no efficient solutions are currently known. The problem can always be simplified, by replacing the entire network or parts of it by their corresponding impulse responses. This can become a disadvantage if parts of the network are frequently updated. Actual applications demand an individual inspection. Stereo processing (including binaural filtering) can be accelerated using complex-data fast Fourier transforms.

8.3. Outlook

The contribution of this thesis is the study of different techniques for real-time filtering and their properties—in other words, the building blocks for auralization systems. Yet, many questions remain open and mark the origin for future research in the field. This accords for elementary tools, the algorithms themselves, but especially their integration in applications. Furthermore, they need to be compared to different filtering concepts. This final section tries do identify these important scientific questions.

First off is the general question of the technique to use: FIR filters or IIR filters. Binaural synthesis for instance can be efficiently realized with infinite impulse response (IIR) head-related transfer function (HRTF) filters of surprisingly little orders. These methods should be compared to FIR approaches based on short linear convolution techniques. Powerful FIR filtering algorithms enable the physically-based auralization with simulated impulse responses. Closed-loop filter structures, e.g. feedback delay networks, reproduce the reverberation characteristic (in time and frequency), but fail to reproduce the fine structure of an impulse response. Recent approaches, like perceptual fast convolution, start moving into the direction of approximate FIR filtering. This raises the question, if the often appreciated exactness of FIR filters might exceed the requirements and is actually necessary. Interesting would be a head-to-head comparison of both classes of filters with respect to the human perception in multi-modal environments, examining the required quality (listening tests) and algorithmic complexity.

The possibility of simulating the reverberation of a space using FIR filters, is mainly reasoned by highly efficient fast convolution techniques, in particular non-uniformly partitioned methods. Few publications address the implementation of the latter on different computing architectures. How to achieve an even load distribution with these algorithms has not been finally clarified. Current concepts for dynamic scheduling the occurring sub convolutions feature a shortest-deadline-first approach, which does not necessarily even out the load curve. The flexibility, inherent in the non-uniform partition, could be further exploited in this respect. Especially the required amount of flexibility should be examined in implementations. Parameters of non-uniform filtering techniques for the application with room acoustic auralization need further research. Open questions are for instance the necessary filter update rates and influence of filter exchange strategies (coherent vs. asynchronous) on the perception in an interactive auditory environment.

8. Conclusion

Finally, after more than 60 years of research, the convolution algorithms themselves still offer room for improvements. A promising concept is the acceleration using multi-dimensional index mapping applied to linear convolution. Uniformly-partitioned running convolutions can be interpreted as a special case of two-dimensional convolutions. From this perspective, most current concepts implement only the larger dimension in a fast way. Hurchalla [47] presented how also the smaller dimension can benefit from using fast short linear convolution techniques. This eventually closes the circle around the methods introduced in chapter 2 and increases the importance of short-length convolution algorithms (e.g. interpolation-based). Not all of these methods confirm with a real-time processing. The computational savings need to be analyzed in more detail and suitable short convolutions have to be found for uniformly-partitioned segments with larger multiplicities in non-uniformly partitioned techniques.

A. Appendix

A.1. Real-time audio processing

Low-latency audio applications on modern computers (e.g. PCs, tablets, mobile devices, game consoles) are realized by block-wise audio streaming. Samples are exchanged with the audio hardware via interrupts (*callbacks*), following the pull-model (audio driver calls application, not the other way round). The audio device delivers and requests input and output samples of a fixed *sampling rate* f_S in units of blocks, with a fixed number of samples B, known as the buffer size or just *block length*. An *audio stream* consists of an indefinite number of length-B blocks (frames) $x^{(i)}(n)$ corresponding to a quasi-continuous signal $x(n)$. In other words, the signal $x(n)$ is uniformly-partitioned into length-B blocks.

Latency

For the signal processing, each recorded block of samples needs to be transported from the audio device (e.g. PCI card) to the central processing unit (CPU) over the machines bus system. After the computations are finished, it is transferred back to the audio device for playback. Note, that also for playback-only applications (e.g. sound synthesis) involve at least this step. These intermediate steps introduce a delay, referred to as *input-to-output latency*, symbolized by ΔT_{IO}. Most bus systems (e.g. PCI, PCI-X) realize data communication between multiple devices using use time-division multiplexing (TDM) with a limited data size per bus transaction. Transmitting and processing individual samples is possible, but would cause a significant overhead on the bus and the intermediate buffering stages. Moreover, a consistent guaranteed bus access in the duration of samples (e.g. 21 μs at 44100 Hz) is difficult to realize. Thus a block-based paradigm is much more favorable in these circumstances.

The *duration* T_B of a block and the *block rate* f_B (number of blocks per seconds) are defined as

$$T_B = \frac{B}{f_S} \ [\text{s}] \qquad f_B = \frac{f_S}{B} \ [\text{s}^{-1}] \qquad T_B = \frac{1}{f_S} \qquad (\text{A.1})$$

Typically a double-buffering strategy is applied for the data exchange with the audio hardware, which introduces an input-to-output latency $\Delta T_{\text{IO}} > 2T_B$ of at least two blocks. Additional latencies are contributed by the hardware. They are significant and should not be neglected (e.g. 2.5 ms or 110 samples at 44100 Hz for AD/DA conversion and transmission for an RME Hammerfall HDSP in a loopback experiment). Table A.1 provides an overview about common block lengths and their corresponding rates and durations.

Block lengths

First, the sampling rate is chosen with respect to the fidelity. A common choice for audio applications is $f_S = 44100$ Hz, covering the audible frequency range of 20-20000 Hz. Secondly, the block length B it usually adjusted to meet the latency requirements of the application. Typical block lengths are small powers of two, e.g. $B \in \{32, 64, 128, 256, 512\}$. Internally within the application, smaller or larger block lengths B' can be used. Smaller block lengths $B' < B$ can be used, as long as they subdivide the original block $B' \mid B$. Then each length-B block is processed by internally iterating over multiple length-B' blocks. However, this approach will not lower the overall latency and usually increase the computational costs, as the succeeding chapters will show. On the other hand, a larger block length lowers the computational load, but requires to accumulate multiple blocks and thus cause additional delays. In order to keep the latency ΔT_{IO} minimal, the samples of the audio stream may not be processed in blocks larger than B samples. Consequently, the first sub filter in a partitioned convolution is commonly implemented using the streaming block length B (see chapters 5, 6). This choice usually marks a fixed constraint. Finally, the latency can be decreased by using a higher sampling rate f_S, but also increases the computational load.

Semantic

The subsequent considerations demand a well-defined semantic of the block-based processing. Generally, samples of a uniform sampling in time align with a quantized time scale $t \in \{ nT + T_0 \mid n \in \mathbb{N} \}$, determined by the sampling rate f_S and block length B. $n = 0, 1, \ldots$ marks the index of a sample and starts with 0. $T = 1/f_S$ is the period length of a sample. T_0 denotes an offset (start time of the stream) and is neglected for the subsequent considerations. If the audio processing is performed block-wise, in blocks of length B, the underlying time scale is $t \in \{ nT_B + T_0 \mid n \in \mathbb{N} \}$. $T_B = B/f_S$ denoted the duration of a length-B block. Audio interrupts (*callbacks*) occur at fixed times $t_n = nT + T_0$. Hence, $k \in \mathbb{N}$ is referred to as the *stream cycle*,

Block length B	Block duration $T_B = B/f_S$ [s]	Block rate $f_B = f_S/B$ [s^{-1}]
32	726 μs	1378.1 s^{-1}
64	1.45 ms	689.1 s^{-1}
128	2.90 ms	344.5 s^{-1}
256	5.80 ms	172.3 s^{-1}
512	11.61 ms	86.1 s^{-1}
1,024	23.22 ms	43.1 s^{-1}
2,048	46.44 ms	21.5 s^{-1}
4,096	92.88 ms	10.7 s^{-1}
8,192	185.76 ms	5.4 s^{-1}
16,384	371.52 ms	2.7 s^{-1}

Table A.1.: Block measures in real-time audio streaming (sampling rate $f_S = 44,100$).

also starting with 0. Again, the start time of the streaming T_0 is not of importance.

Entering the first cycle $(k = 0)$, the audio driver provides the first length-B block of recorded input samples $x_0(n)$ and requests the first length-B block $y_0(n)$ of output samples for playback. In the first cycle $k = 0$ for example, the input samples $x^{(0)}(0), \ldots, x^{(0)}(B-1)$ are provided and the output samples $y^{(0)}(0), \ldots, y^{(0)}(B-1)$ are requested. Accordingly, with in the cycle k, samples with indices n in the range $kB \le n < (k+1)B$ are processed. The cycle k in which the sample of index n falls into is given by

$$k = \left\lfloor \frac{n}{B} \right\rfloor \tag{A.2}$$

The signal processing in a cycle must be finished before the next cycle begins. The maximum computation time in each cycle is defined as $T_{\max}$ $T_{\max} = T_B - T_e$. It may not exceed the duration T_B of a block. Call overheads and intermediate buffering consume additional time and require a safety margin T_e which depends on the hardware and software. Typically, a stable operation is possible with a margins in the range of 5-10%, meaning that about 90-95% of the block duration T_B can be spent on signal processing. If the time budget $T_{\max}$ is exceeded, then the audio device is not provided with the next output samples in time, causing dropouts in the playback.

A.2. Optimized non-uniform filter partitions

The following tables lists optimal partitions for the standard NUPOLS algorithm on the test system. The optimization founds on benchmarks of all involved operations described in chapter 7. The computational costs are evaluated using the model in Eq. 6.12. The set of optimal filter partitions was determined using the Viterbi algorithm, as described in section 6.5. The results here can differ from those in prior publications [121], which found on other sets of performance data. The first column is the desired target filter length, followed by the computational costs of the optimal partition. The costs are measured as (average) CPU cycles per filter output sample. In the next column the actual optimal partition is written. The last column lists the clearances (block-multiples) for all segments in this partition. The lists are complete and cover all optimal partitions for all filter lengths $B \leq N \leq 2^{20}$ in steps of B samples. Given that two filter length $N' < N$ share the same optimal partition, the length N' is skipped in the table.

A.2.1. Minimal-load partitions for the block length B=128

Filter length N	Cost/sample	Optimal partition $\mathcal{P}_{\text{opt}}$	Segment clearances
128	43.0	(128^1)	(0)
256	45.8	(128^2)	(0)
384	48.7	(128^3)	(0)
512	51.6	(128^4)	(0)
640	54.4	(128^5)	(0)
768	57.3	(128^6)	(0)
896	60.2	(128^7)	(0)
1024	63.0	(128^8)	(0)
1152	65.9	(128^9)	(0)
1280	68.8	(128^{10})	(0)
1408	71.6	(128^{11})	(0)
1536	74.5	(128^{12})	(0)
1664	77.4	(128^{13})	(0)
1792	80.2	(128^{14})	(0)
1920	83.1	(128^{15})	(0)
2048	86.0	(128^{16})	(0)
2176	88.8	(128^{17})	(0)
2304	91.7	(128^{18})	(0)
2432	94.6	(128^{19})	(0)
2560	97.4	(128^{20})	(0)
2944	100.1	$(128^3\, 512^5)$	$(0\ 0)$
3456	102.7	$(128^3\, 512^6)$	$(0\ 0)$
3968	105.3	$(128^3\, 512^7)$	$(0\ 0)$
4480	108.0	$(128^3\, 512^8)$	$(0\ 0)$

4992	110.6	$(128^3\, 512^9)$	$(0\ 0)$
5504	113.2	$(128^3\, 512^{10})$	$(0\ 0)$
6016	113.5	$(128^7\, 1024^5)$	$(0\ 0)$
7040	116.1	$(128^7\, 1024^6)$	$(0\ 0)$
8064	118.7	$(128^7\, 1024^7)$	$(0\ 0)$
9088	121.2	$(128^7\, 1024^8)$	$(0\ 0)$
10112	123.8	$(128^7\, 1024^9)$	$(0\ 0)$
11136	126.3	$(128^7\, 1024^{10})$	$(0\ 0)$
12160	128.9	$(128^7\, 1024^{11})$	$(0\ 0)$
13184	131.5	$(128^7\, 1024^{12})$	$(0\ 0)$
14208	134.0	$(128^7\, 1024^{13})$	$(0\ 0)$
15232	136.6	$(128^7\, 1024^{14})$	$(0\ 0)$
16256	139.1	$(128^7\, 1024^{15})$	$(0\ 0)$
17280	141.7	$(128^7\, 1024^{16})$	$(0\ 0)$
18304	144.3	$(128^7\, 1024^{17})$	$(0\ 0)$
19328	146.8	$(128^7\, 1024^{18})$	$(0\ 0)$
20352	149.4	$(128^7\, 1024^{19})$	$(0\ 0)$
21376	152.0	$(128^7\, 1024^{20})$	$(0\ 0)$
22400	154.5	$(128^7\, 1024^{21})$	$(0\ 0)$
23424	157.1	$(128^7\, 1024^{22})$	$(0\ 0)$
24448	159.6	$(128^7\, 1024^{23})$	$(0\ 0)$
25472	162.2	$(128^7\, 1024^{24})$	$(0\ 0)$
26496	164.8	$(128^7\, 1024^{25})$	$(0\ 0)$
27520	167.3	$(128^7\, 1024^{26})$	$(0\ 0)$
28544	169.9	$(128^7\, 1024^{27})$	$(0\ 0)$
29568	172.4	$(128^7\, 1024^{28})$	$(0\ 0)$
30592	175.0	$(128^7\, 1024^{29})$	$(0\ 0)$
31616	177.6	$(128^7\, 1024^{30})$	$(0\ 0)$
32640	180.1	$(128^7\, 1024^{31})$	$(0\ 0)$
33664	182.7	$(128^7\, 1024^{32})$	$(0\ 0)$
34688	183.7	$(128^{15}\, 2048^{16})$	$(0\ 0)$
40832	183.8	$(128^7\, 1024^7\, 8192^4)$	$(0\ 0\ 0)$
41856	186.4	$(128^7\, 1024^8\, 8192^4)$	$(0\ 0\ 8)$
49024	187.2	$(128^7\, 1024^7\, 8192^5)$	$(0\ 0\ 0)$
50048	189.8	$(128^7\, 1024^8\, 8192^5)$	$(0\ 0\ 8)$
57216	190.6	$(128^7\, 1024^7\, 8192^6)$	$(0\ 0\ 0)$
58240	193.2	$(128^7\, 1024^8\, 8192^6)$	$(0\ 0\ 8)$
65408	194.0	$(128^7\, 1024^7\, 8192^7)$	$(0\ 0\ 0)$
66432	196.5	$(128^7\, 1024^8\, 8192^7)$	$(0\ 0\ 8)$
73600	197.4	$(128^7\, 1024^7\, 8192^8)$	$(0\ 0\ 0)$
74624	199.9	$(128^7\, 1024^8\, 8192^8)$	$(0\ 0\ 8)$
81792	200.7	$(128^7\, 1024^7\, 8192^9)$	$(0\ 0\ 0)$
82816	203.3	$(128^7\, 1024^8\, 8192^9)$	$(0\ 0\ 8)$
89984	204.1	$(128^7\, 1024^7\, 8192^{10})$	$(0\ 0\ 0)$
91008	206.7	$(128^7\, 1024^8\, 8192^{10})$	$(0\ 0\ 8)$
98176	207.5	$(128^7\, 1024^7\, 8192^{11})$	$(0\ 0\ 0)$
99200	210.1	$(128^7\, 1024^8\, 8192^{11})$	$(0\ 0\ 8)$
106368	210.9	$(128^7\, 1024^7\, 8192^{12})$	$(0\ 0\ 0)$
107392	213.5	$(128^7\, 1024^8\, 8192^{12})$	$(0\ 0\ 8)$
114560	214.3	$(128^7\, 1024^7\, 8192^{13})$	$(0\ 0\ 0)$
115584	216.8	$(128^7\, 1024^8\, 8192^{13})$	$(0\ 0\ 8)$

122752	217.7	(128^7 1024^7 8192^{14})	(0 0 0)
123776	220.2	(128^7 1024^8 8192^{14})	(0 0 8)
130944	220.9	(128^7 1024^{15} 16384^7)	(0 0 0)
131968	223.4	(128^7 1024^{16} 16384^7)	(0 0 8)
139136	224.4	(128^7 1024^7 8192^{16})	(0 0 0)
147328	224.6	(128^7 1024^{15} 16384^8)	(0 0 0)
148352	227.1	(128^7 1024^{16} 16384^8)	(0 0 8)
163712	228.3	(128^7 1024^{15} 16384^9)	(0 0 0)
164736	230.8	(128^7 1024^{16} 16384^9)	(0 0 8)
180096	232.0	(128^7 1024^{15} 16384^{10})	(0 0 0)
181120	234.6	(128^7 1024^{16} 16384^{10})	(0 0 8)
196480	235.7	(128^7 1024^{15} 16384^{11})	(0 0 0)
197504	238.3	(128^7 1024^{16} 16384^{11})	(0 0 8)
212864	239.4	(128^7 1024^{15} 16384^{12})	(0 0 0)
213888	242.0	(128^7 1024^{16} 16384^{12})	(0 0 8)
229248	243.1	(128^7 1024^{15} 16384^{13})	(0 0 0)
230272	245.7	(128^7 1024^{16} 16384^{13})	(0 0 8)
245632	246.8	(128^7 1024^{15} 16384^{14})	(0 0 0)
246656	249.4	(128^7 1024^{16} 16384^{14})	(0 0 8)
262016	250.6	(128^7 1024^{15} 16384^{15})	(0 0 0)
263040	253.1	(128^7 1024^{16} 16384^{15})	(0 0 8)
278400	254.3	(128^7 1024^{15} 16384^{16})	(0 0 0)
279424	256.8	(128^7 1024^{16} 16384^{16})	(0 0 8)
294784	258.0	(128^7 1024^{15} 16384^{17})	(0 0 0)
295808	260.5	(128^7 1024^{16} 16384^{17})	(0 0 8)
311168	261.7	(128^7 1024^{15} 16384^{18})	(0 0 0)
312192	264.2	(128^7 1024^{16} 16384^{18})	(0 0 8)
327552	265.4	(128^7 1024^{15} 16384^{19})	(0 0 0)
328576	268.0	(128^7 1024^{16} 16384^{19})	(0 0 8)
343936	269.1	(128^7 1024^{15} 16384^{20})	(0 0 0)
344960	271.7	(128^7 1024^{16} 16384^{20})	(0 0 8)
360320	272.8	(128^7 1024^{15} 16384^{21})	(0 0 0)
361344	275.4	(128^7 1024^{16} 16384^{21})	(0 0 8)
376704	276.5	(128^7 1024^{15} 16384^{22})	(0 0 0)
393088	277.3	(128^7 1024^7 8192^7 65536^5)	(0 0 0 0)
394112	279.9	(128^7 1024^8 8192^7 65536^5)	(0 0 8 8)
401280	280.7	(128^7 1024^7 8192^8 65536^5)	(0 0 0 64)
458624	281.1	(128^7 1024^7 8192^7 65536^6)	(0 0 0 0)
459648	283.6	(128^7 1024^8 8192^7 65536^6)	(0 0 8 8)
466816	284.4	(128^7 1024^7 8192^8 65536^6)	(0 0 0 64)
524160	284.8	(128^7 1024^7 8192^7 65536^7)	(0 0 0 0)

A.2.2. Minimal-load partitions for the block length B=256

Filter length N	Cost/sample	Optimal partition $\mathcal{P}_{\text{opt}}$	Segment clearances
256	40.0	(256^1)	(0)
512	42.7	(256^2)	(0)
768	45.4	(256^3)	(0)
1024	48.1	(256^4)	(0)
1280	50.7	(256^5)	(0)
1536	53.4	(256^6)	(0)
1792	56.1	(256^7)	(0)
2048	58.8	(256^8)	(0)
2304	61.5	(256^9)	(0)
2560	64.2	(256^{10})	(0)
2816	66.9	(256^{11})	(0)
3072	69.5	(256^{12})	(0)
3328	72.2	(256^{13})	(0)
3584	74.9	(256^{14})	(0)
3840	77.6	(256^{15})	(0)
4096	80.3	(256^{16})	(0)
4352	83.0	(256^{17})	(0)
4608	85.7	(256^{18})	(0)
4864	88.4	(256^{19})	(0)
5120	91.0	(256^{20})	(0)
5376	93.7	(256^{21})	(0)
5632	96.4	(256^{22})	(0)
5888	98.7	($256^3\ 1024^5$)	(0 0)
6912	101.3	($256^3\ 1024^6$)	(0 0)
7936	103.9	($256^3\ 1024^7$)	(0 0)
8960	106.4	($256^3\ 1024^8$)	(0 0)
9984	109.0	($256^3\ 1024^9$)	(0 0)
11008	111.6	($256^3\ 1024^{10}$)	(0 0)
12032	114.1	($256^3\ 1024^{11}$)	(0 0)
13056	116.7	($256^3\ 1024^{12}$)	(0 0)
14080	119.2	($256^3\ 1024^{13}$)	(0 0)
15104	121.8	($256^3\ 1024^{14}$)	(0 0)
16128	124.4	($256^3\ 1024^{15}$)	(0 0)
17152	126.9	($256^3\ 1024^{16}$)	(0 0)
18176	129.5	($256^3\ 1024^{17}$)	(0 0)
19200	132.0	($256^3\ 1024^{18}$)	(0 0)
20224	133.2	($256^7\ 2048^9$)	(0 0)
20480	135.9	($256^8\ 2048^9$)	(0 1)
22272	136.6	($256^7\ 2048^{10}$)	(0 0)
22528	139.2	($256^8\ 2048^{10}$)	(0 1)
24320	139.9	($256^7\ 2048^{11}$)	(0 0)
24576	142.6	($256^8\ 2048^{11}$)	(0 1)
26368	143.3	($256^7\ 2048^{12}$)	(0 0)
26624	145.9	($256^8\ 2048^{12}$)	(0 1)
28416	146.6	($256^7\ 2048^{13}$)	(0 0)

28672	149.3	$(256^8\ 2048^{13})$	$(0\ 1)$
30464	150.0	$(256^7\ 2048^{14})$	$(0\ 0)$
30720	152.6	$(256^8\ 2048^{14})$	$(0\ 1)$
32512	152.8	$(256^{15}\ 4096^7)$	$(0\ 0)$
32768	155.5	$(256^{16}\ 4096^7)$	$(0\ 1)$
36608	156.2	$(256^{15}\ 4096^8)$	$(0\ 0)$
36864	158.9	$(256^{16}\ 4096^8)$	$(0\ 1)$
40704	159.6	$(256^{15}\ 4096^9)$	$(0\ 0)$
40960	162.3	$(256^{16}\ 4096^9)$	$(0\ 1)$
44800	163.0	$(256^{15}\ 4096^{10})$	$(0\ 0)$
45056	165.7	$(256^{16}\ 4096^{10})$	$(0\ 1)$
48896	166.4	$(256^{15}\ 4096^{11})$	$(0\ 0)$
49152	169.0	$(256^{16}\ 4096^{11})$	$(0\ 1)$
52992	169.7	$(256^{15}\ 4096^{12})$	$(0\ 0)$
53248	172.4	$(256^{16}\ 4096^{12})$	$(0\ 1)$
57088	173.1	$(256^{15}\ 4096^{13})$	$(0\ 0)$
57344	175.8	$(256^{16}\ 4096^{13})$	$(0\ 1)$
61184	176.5	$(256^{15}\ 4096^{14})$	$(0\ 0)$
65280	179.2	$(256^3\ 1024^7\ 8192^7)$	$(0\ 0\ 0)$
66304	181.7	$(256^3\ 1024^8\ 8192^7)$	$(0\ 0\ 4)$
73472	182.6	$(256^3\ 1024^7\ 8192^8)$	$(0\ 0\ 0)$
74496	185.1	$(256^3\ 1024^8\ 8192^8)$	$(0\ 0\ 4)$
81664	186.0	$(256^3\ 1024^7\ 8192^9)$	$(0\ 0\ 0)$
82688	188.5	$(256^3\ 1024^8\ 8192^9)$	$(0\ 0\ 4)$
89856	189.3	$(256^3\ 1024^7\ 8192^{10})$	$(0\ 0\ 0)$
90880	191.9	$(256^3\ 1024^8\ 8192^{10})$	$(0\ 0\ 4)$
98048	192.7	$(256^3\ 1024^7\ 8192^{11})$	$(0\ 0\ 0)$
99072	195.3	$(256^3\ 1024^8\ 8192^{11})$	$(0\ 0\ 4)$
106240	196.1	$(256^3\ 1024^7\ 8192^{12})$	$(0\ 0\ 0)$
107264	198.7	$(256^3\ 1024^8\ 8192^{12})$	$(0\ 0\ 4)$
114432	199.5	$(256^3\ 1024^7\ 8192^{13})$	$(0\ 0\ 0)$
115456	202.0	$(256^3\ 1024^8\ 8192^{13})$	$(0\ 0\ 4)$
122624	202.9	$(256^3\ 1024^7\ 8192^{14})$	$(0\ 0\ 0)$
123648	205.4	$(256^3\ 1024^8\ 8192^{14})$	$(0\ 0\ 4)$
130816	206.1	$(256^3\ 1024^{15}\ 16384^7)$	$(0\ 0\ 0)$
131840	208.6	$(256^3\ 1024^{16}\ 16384^7)$	$(0\ 0\ 4)$
139008	209.6	$(256^3\ 1024^7\ 8192^{16})$	$(0\ 0\ 0)$
147200	209.8	$(256^3\ 1024^{15}\ 16384^8)$	$(0\ 0\ 0)$
148224	212.3	$(256^3\ 1024^{16}\ 16384^8)$	$(0\ 0\ 4)$
163584	213.5	$(256^3\ 1024^{15}\ 16384^9)$	$(0\ 0\ 0)$
164608	216.1	$(256^3\ 1024^{16}\ 16384^9)$	$(0\ 0\ 4)$
179968	217.2	$(256^3\ 1024^{15}\ 16384^{10})$	$(0\ 0\ 0)$
180992	219.8	$(256^3\ 1024^{16}\ 16384^{10})$	$(0\ 0\ 4)$
196352	220.9	$(256^3\ 1024^{15}\ 16384^{11})$	$(0\ 0\ 0)$
197376	223.5	$(256^3\ 1024^{16}\ 16384^{11})$	$(0\ 0\ 4)$
212736	224.6	$(256^3\ 1024^{15}\ 16384^{12})$	$(0\ 0\ 0)$
213760	227.2	$(256^3\ 1024^{16}\ 16384^{12})$	$(0\ 0\ 4)$
229120	228.3	$(256^3\ 1024^{15}\ 16384^{13})$	$(0\ 0\ 0)$
230144	230.9	$(256^3\ 1024^{16}\ 16384^{13})$	$(0\ 0\ 4)$
245504	232.0	$(256^3\ 1024^{15}\ 16384^{14})$	$(0\ 0\ 0)$
246528	234.6	$(256^3\ 1024^{16}\ 16384^{14})$	$(0\ 0\ 4)$

261888	235.8	$(\, 256^3 \, 1024^{15} \, 16384^{15} \,)$	$(\, 0 \; 0 \; 0 \,)$
262912	238.3	$(\, 256^3 \, 1024^{16} \, 16384^{15} \,)$	$(\, 0 \; 0 \; 4 \,)$
278272	239.5	$(\, 256^3 \, 1024^{15} \, 16384^{16} \,)$	$(\, 0 \; 0 \; 0 \,)$
294656	241.5	$(\, 256^{15} \, 4096^7 \, 32768^8 \,)$	$(\, 0 \; 0 \; 0 \,)$
294912	244.2	$(\, 256^{16} \, 4096^7 \, 32768^8 \,)$	$(\, 0 \; 1 \; 1 \,)$
298752	244.9	$(\, 256^{15} \, 4096^8 \, 32768^8 \,)$	$(\, 0 \; 0 \; 16 \,)$
327424	245.2	$(\, 256^{15} \, 4096^7 \, 32768^9 \,)$	$(\, 0 \; 0 \; 0 \,)$
327680	247.9	$(\, 256^{16} \, 4096^7 \, 32768^9 \,)$	$(\, 0 \; 1 \; 1 \,)$
331520	248.6	$(\, 256^{15} \, 4096^8 \, 32768^9 \,)$	$(\, 0 \; 0 \; 16 \,)$
360192	248.9	$(\, 256^{15} \, 4096^7 \, 32768^{10} \,)$	$(\, 0 \; 0 \; 0 \,)$
360448	251.6	$(\, 256^{16} \, 4096^7 \, 32768^{10} \,)$	$(\, 0 \; 1 \; 1 \,)$
364288	252.3	$(\, 256^{15} \, 4096^8 \, 32768^{10} \,)$	$(\, 0 \; 0 \; 16 \,)$
392960	252.6	$(\, 256^{15} \, 4096^7 \, 32768^{11} \,)$	$(\, 0 \; 0 \; 0 \,)$
393216	255.3	$(\, 256^{16} \, 4096^7 \, 32768^{11} \,)$	$(\, 0 \; 1 \; 1 \,)$
397056	256.0	$(\, 256^{15} \, 4096^8 \, 32768^{11} \,)$	$(\, 0 \; 0 \; 16 \,)$
425728	256.3	$(\, 256^{15} \, 4096^7 \, 32768^{12} \,)$	$(\, 0 \; 0 \; 0 \,)$
425984	259.0	$(\, 256^{16} \, 4096^7 \, 32768^{12} \,)$	$(\, 0 \; 1 \; 1 \,)$
429824	259.7	$(\, 256^{15} \, 4096^8 \, 32768^{12} \,)$	$(\, 0 \; 0 \; 16 \,)$
458496	260.0	$(\, 256^{15} \, 4096^7 \, 32768^{13} \,)$	$(\, 0 \; 0 \; 0 \,)$
458752	262.7	$(\, 256^{16} \, 4096^7 \, 32768^{13} \,)$	$(\, 0 \; 1 \; 1 \,)$
462592	263.4	$(\, 256^{15} \, 4096^8 \, 32768^{13} \,)$	$(\, 0 \; 0 \; 16 \,)$
491264	263.7	$(\, 256^{15} \, 4096^7 \, 32768^{14} \,)$	$(\, 0 \; 0 \; 0 \,)$
491520	266.4	$(\, 256^{16} \, 4096^7 \, 32768^{14} \,)$	$(\, 0 \; 1 \; 1 \,)$
495360	267.1	$(\, 256^{15} \, 4096^8 \, 32768^{14} \,)$	$(\, 0 \; 0 \; 16 \,)$
524032	267.4	$(\, 256^{15} \, 4096^7 \, 32768^{15} \,)$	$(\, 0 \; 0 \; 0 \,)$
524288	270.1	$(\, 256^{16} \, 4096^7 \, 32768^{15} \,)$	$(\, 0 \; 1 \; 1 \,)$

A.2.3. Minimal-load partitions for the block length B=512

Filter length N	Cost/sample	Optimal partition $\mathcal{P}_{opt}$	Segment clearances
512	40.9	$(\, 512^1 \,)$	$(\, 0 \,)$
1024	43.5	$(\, 512^2 \,)$	$(\, 0 \,)$
1536	46.2	$(\, 512^3 \,)$	$(\, 0 \,)$
2048	48.8	$(\, 512^4 \,)$	$(\, 0 \,)$
2560	51.4	$(\, 512^5 \,)$	$(\, 0 \,)$
3072	54.0	$(\, 512^6 \,)$	$(\, 0 \,)$
3584	56.6	$(\, 512^7 \,)$	$(\, 0 \,)$
4096	59.3	$(\, 512^8 \,)$	$(\, 0 \,)$
4608	61.9	$(\, 512^9 \,)$	$(\, 0 \,)$
5120	64.5	$(\, 512^{10} \,)$	$(\, 0 \,)$
5632	67.1	$(\, 512^{11} \,)$	$(\, 0 \,)$
6144	69.8	$(\, 512^{12} \,)$	$(\, 0 \,)$
6656	72.4	$(\, 512^{13} \,)$	$(\, 0 \,)$
7168	75.0	$(\, 512^{14} \,)$	$(\, 0 \,)$
7680	77.6	$(\, 512^{15} \,)$	$(\, 0 \,)$
8192	80.3	$(\, 512^{16} \,)$	$(\, 0 \,)$
8704	82.9	$(\, 512^{17} \,)$	$(\, 0 \,)$

9216	85.5	(512^{18})	(0)
9728	88.1	(512^{19})	(0)
10240	90.7	(512^{20})	(0)
10752	93.4	(512^{21})	(0)
11264	96.0	(512^{22})	(0)
11776	98.6	(512^{23})	(0)
12288	101.2	(512^{24})	(0)
12800	103.9	(512^{25})	(0)
13312	106.5	(512^{26})	(0)
13824	109.1	(512^{27})	(0)
14336	111.7	(512^{28})	(0)
14848	114.4	(512^{29})	(0)
15872	116.5	$(512^3\,2048^7)$	$(0\ 0)$
16384	119.2	$(512^4\,2048^7)$	$(0\ 1)$
17920	119.9	$(512^3\,2048^8)$	$(0\ 0)$
19968	121.7	$(512^7\,4096^4)$	$(0\ 0)$
20480	124.4	$(512^8\,4096^4)$	$(0\ 1)$
24064	125.1	$(512^7\,4096^5)$	$(0\ 0)$
24576	127.7	$(512^8\,4096^5)$	$(0\ 1)$
28160	128.5	$(512^7\,4096^6)$	$(0\ 0)$
28672	131.1	$(512^8\,4096^6)$	$(0\ 1)$
32256	131.9	$(512^7\,4096^7)$	$(0\ 0)$
32768	134.5	$(512^8\,4096^7)$	$(0\ 1)$
36352	135.3	$(512^7\,4096^8)$	$(0\ 0)$
36864	137.9	$(512^8\,4096^8)$	$(0\ 1)$
40448	138.6	$(512^7\,4096^9)$	$(0\ 0)$
40960	141.3	$(512^8\,4096^9)$	$(0\ 1)$
44544	142.0	$(512^7\,4096^{10})$	$(0\ 0)$
45056	144.6	$(512^8\,4096^{10})$	$(0\ 1)$
48640	145.4	$(512^7\,4096^{11})$	$(0\ 0)$
49152	148.0	$(512^8\,4096^{11})$	$(0\ 1)$
52736	148.8	$(512^7\,4096^{12})$	$(0\ 0)$
56832	149.6	$(512^{15}\,8192^6)$	$(0\ 0)$
57344	152.2	$(512^{16}\,8192^6)$	$(0\ 1)$
65024	152.9	$(512^{15}\,8192^7)$	$(0\ 0)$
65536	155.6	$(512^{16}\,8192^7)$	$(0\ 1)$
73216	156.3	$(512^{15}\,8192^8)$	$(0\ 0)$
73728	159.0	$(512^{16}\,8192^8)$	$(0\ 1)$
81408	159.7	$(512^{15}\,8192^9)$	$(0\ 0)$
81920	162.3	$(512^{16}\,8192^9)$	$(0\ 1)$
89600	163.1	$(512^{15}\,8192^{10})$	$(0\ 0)$
90112	165.7	$(512^{16}\,8192^{10})$	$(0\ 1)$
97792	166.5	$(512^{15}\,8192^{11})$	$(0\ 0)$
98304	169.1	$(512^{16}\,8192^{11})$	$(0\ 1)$
105984	169.9	$(512^{15}\,8192^{12})$	$(0\ 0)$
106496	172.5	$(512^{16}\,8192^{12})$	$(0\ 1)$
114176	173.2	$(512^{15}\,8192^{13})$	$(0\ 0)$
114688	175.9	$(512^{16}\,8192^{13})$	$(0\ 1)$
122368	176.6	$(512^{15}\,8192^{14})$	$(0\ 0)$
122880	179.3	$(512^{16}\,8192^{14})$	$(0\ 1)$
130560	180.0	$(512^{15}\,8192^{15})$	$(0\ 0)$

131072	182.6	$(\,512^{16}\,8192^{15}\,)$	$(\,0\ 1\,)$
138752	183.4	$(\,512^{15}\,8192^{16}\,)$	$(\,0\ 0\,)$
139264	186.0	$(\,512^{16}\,8192^{16}\,)$	$(\,0\ 1\,)$
146944	186.8	$(\,512^{15}\,8192^{17}\,)$	$(\,0\ 0\,)$
147456	189.4	$(\,512^{16}\,8192^{17}\,)$	$(\,0\ 1\,)$
155136	190.2	$(\,512^{15}\,8192^{18}\,)$	$(\,0\ 0\,)$
155648	192.8	$(\,512^{16}\,8192^{18}\,)$	$(\,0\ 1\,)$
163328	193.6	$(\,512^{15}\,8192^{19}\,)$	$(\,0\ 0\,)$
163840	196.2	$(\,512^{16}\,8192^{19}\,)$	$(\,0\ 1\,)$
171520	196.9	$(\,512^{15}\,8192^{20}\,)$	$(\,0\ 0\,)$
172032	199.6	$(\,512^{16}\,8192^{20}\,)$	$(\,0\ 1\,)$
179712	200.3	$(\,512^{15}\,8192^{21}\,)$	$(\,0\ 0\,)$
180224	202.9	$(\,512^{16}\,8192^{21}\,)$	$(\,0\ 1\,)$
187904	203.7	$(\,512^{15}\,8192^{22}\,)$	$(\,0\ 0\,)$
188416	206.3	$(\,512^{16}\,8192^{22}\,)$	$(\,0\ 1\,)$
196096	207.1	$(\,512^{15}\,8192^{23}\,)$	$(\,0\ 0\,)$
196608	209.7	$(\,512^{16}\,8192^{23}\,)$	$(\,0\ 1\,)$
204288	210.5	$(\,512^{15}\,8192^{24}\,)$	$(\,0\ 0\,)$
204800	213.1	$(\,512^{16}\,8192^{24}\,)$	$(\,0\ 1\,)$
228864	213.1	$(\,512^{7}\,4096^{7}\,32768^{6}\,)$	$(\,0\ 0\ 0\,)$
229376	215.7	$(\,512^{8}\,4096^{7}\,32768^{6}\,)$	$(\,0\ 1\ 1\,)$
232960	216.5	$(\,512^{7}\,4096^{8}\,32768^{6}\,)$	$(\,0\ 0\ 8\,)$
261632	216.8	$(\,512^{7}\,4096^{7}\,32768^{7}\,)$	$(\,0\ 0\ 0\,)$
262144	219.4	$(\,512^{8}\,4096^{7}\,32768^{7}\,)$	$(\,0\ 1\ 1\,)$
265728	220.2	$(\,512^{7}\,4096^{8}\,32768^{7}\,)$	$(\,0\ 0\ 8\,)$
294400	220.5	$(\,512^{7}\,4096^{7}\,32768^{8}\,)$	$(\,0\ 0\ 0\,)$
294912	223.1	$(\,512^{8}\,4096^{7}\,32768^{8}\,)$	$(\,0\ 1\ 1\,)$
298496	223.9	$(\,512^{7}\,4096^{8}\,32768^{8}\,)$	$(\,0\ 0\ 8\,)$
327168	224.2	$(\,512^{7}\,4096^{7}\,32768^{9}\,)$	$(\,0\ 0\ 0\,)$
327680	226.8	$(\,512^{8}\,4096^{7}\,32768^{9}\,)$	$(\,0\ 1\ 1\,)$
331264	227.6	$(\,512^{7}\,4096^{8}\,32768^{9}\,)$	$(\,0\ 0\ 8\,)$
359936	227.9	$(\,512^{7}\,4096^{7}\,32768^{10}\,)$	$(\,0\ 0\ 0\,)$
360448	230.5	$(\,512^{8}\,4096^{7}\,32768^{10}\,)$	$(\,0\ 1\ 1\,)$
364032	231.3	$(\,512^{7}\,4096^{8}\,32768^{10}\,)$	$(\,0\ 0\ 8\,)$
392704	231.6	$(\,512^{7}\,4096^{7}\,32768^{11}\,)$	$(\,0\ 0\ 0\,)$
393216	234.3	$(\,512^{8}\,4096^{7}\,32768^{11}\,)$	$(\,0\ 1\ 1\,)$
396800	235.0	$(\,512^{7}\,4096^{8}\,32768^{11}\,)$	$(\,0\ 0\ 8\,)$
425472	235.3	$(\,512^{7}\,4096^{7}\,32768^{12}\,)$	$(\,0\ 0\ 0\,)$
425984	238.0	$(\,512^{8}\,4096^{7}\,32768^{12}\,)$	$(\,0\ 1\ 1\,)$
429568	238.7	$(\,512^{7}\,4096^{8}\,32768^{12}\,)$	$(\,0\ 0\ 8\,)$
458240	239.0	$(\,512^{7}\,4096^{7}\,32768^{13}\,)$	$(\,0\ 0\ 0\,)$
458752	241.7	$(\,512^{8}\,4096^{7}\,32768^{13}\,)$	$(\,0\ 1\ 1\,)$
462336	242.4	$(\,512^{7}\,4096^{8}\,32768^{13}\,)$	$(\,0\ 0\ 8\,)$
491008	242.8	$(\,512^{7}\,4096^{7}\,32768^{14}\,)$	$(\,0\ 0\ 0\,)$
523776	243.8	$(\,512^{15}\,8192^{7}\,65536^{7}\,)$	$(\,0\ 0\ 0\,)$
524288	246.4	$(\,512^{16}\,8192^{7}\,65536^{7}\,)$	$(\,0\ 1\ 1\,)$

A.2.4. Practical partitions of variant 1 for the block length B=128

Filter length N	Cost/sample	Optimal partition $\mathcal{P}_{\text{opt}}$	Segment clearances
128	43.0	(128^1)	(0)
256	45.8	(128^2)	(0)
384	48.7	(128^3)	(0)
512	51.6	(128^4)	(0)
640	54.4	(128^5)	(0)
768	57.3	(128^6)	(0)
896	60.2	(128^7)	(0)
1024	63.0	(128^8)	(0)
1152	65.9	(128^9)	(0)
1280	68.8	(128^{10})	(0)
1408	71.6	(128^{11})	(0)
1536	74.5	(128^{12})	(0)
1664	77.4	(128^{13})	(0)
1792	80.2	(128^{14})	(0)
1920	83.1	(128^{15})	(0)
2048	86.0	(128^{16})	(0)
2176	88.8	(128^{17})	(0)
2304	91.7	(128^{18})	(0)
2432	94.6	(128^{19})	(0)
2560	97.4	(128^{20})	(0)
2688	100.3	(128^{21})	(0)
2816	103.2	(128^{22})	(0)
2944	106.0	(128^{23})	(0)
3072	108.9	(128^{24})	(0)
3200	111.8	(128^{25})	(0)
3328	114.7	(128^{26})	(0)
3456	117.5	(128^{27})	(0)
3840	120.2	($128^{10}\,512^5$)	(0 7)
4352	122.8	($128^{10}\,512^6$)	(0 7)
4864	125.4	($128^{10}\,512^7$)	(0 7)
5376	128.0	($128^{10}\,512^8$)	(0 7)
5888	130.7	($128^{10}\,512^9$)	(0 7)
6400	133.3	($128^{10}\,512^{10}$)	(0 7)
6912	135.9	($128^{10}\,512^{11}$)	(0 7)
7424	138.5	($128^{10}\,512^{12}$)	(0 7)
7936	141.1	($128^{10}\,512^{13}$)	(0 7)
8448	143.8	($128^{10}\,512^{14}$)	(0 7)
8960	146.4	($128^{10}\,512^{15}$)	(0 7)
9472	149.0	($128^{10}\,512^{16}$)	(0 7)
9984	151.6	($128^{10}\,512^{17}$)	(0 7)
10496	154.3	($128^{10}\,512^{18}$)	(0 7)
11008	156.9	($128^{10}\,512^{19}$)	(0 7)
11520	159.5	($128^{10}\,512^{20}$)	(0 7)
12032	162.1	($128^{10}\,512^{21}$)	(0 7)
12544	164.8	($128^{10}\,512^{22}$)	(0 7)

13056	167.4	$(128^{10}\, 512^{23})$	(0 7)
13568	170.0	$(128^{10}\, 512^{24})$	(0 7)
14080	171.9	$(128^{22}\, 1024^{11})$	(0 15)
15104	174.5	$(128^{22}\, 1024^{12})$	(0 15)
16128	177.0	$(128^{22}\, 1024^{13})$	(0 15)
17152	179.6	$(128^{22}\, 1024^{14})$	(0 15)
18176	182.2	$(128^{22}\, 1024^{15})$	(0 15)
19200	184.7	$(128^{22}\, 1024^{16})$	(0 15)
20224	187.3	$(128^{22}\, 1024^{17})$	(0 15)
21248	189.9	$(128^{22}\, 1024^{18})$	(0 15)
22272	192.4	$(128^{22}\, 1024^{19})$	(0 15)
23296	195.0	$(128^{22}\, 1024^{20})$	(0 15)
24320	197.5	$(128^{22}\, 1024^{21})$	(0 15)
25344	200.1	$(128^{22}\, 1024^{22})$	(0 15)
26368	202.7	$(128^{22}\, 1024^{23})$	(0 15)
27392	205.2	$(128^{22}\, 1024^{24})$	(0 15)
28416	207.8	$(128^{22}\, 1024^{25})$	(0 15)
29440	210.3	$(128^{22}\, 1024^{26})$	(0 15)
30464	212.9	$(128^{22}\, 1024^{27})$	(0 15)
31488	215.5	$(128^{22}\, 1024^{28})$	(0 15)
32512	218.0	$(128^{22}\, 1024^{29})$	(0 15)
33536	220.6	$(128^{22}\, 1024^{30})$	(0 15)
34560	223.1	$(128^{22}\, 1024^{31})$	(0 15)
35584	225.7	$(128^{22}\, 1024^{32})$	(0 15)
36608	227.9	$(128^{10}\, 512^{9}\, 2048^{15})$	(0 7 31)
37120	230.5	$(128^{10}\, 512^{10}\, 2048^{15})$	(0 7 35)
37632	230.8	$(128^{22}\, 1024^{34})$	(0 15)
38656	231.2	$(128^{10}\, 512^{9}\, 2048^{16})$	(0 7 31)
39168	233.8	$(128^{10}\, 512^{10}\, 2048^{16})$	(0 7 35)
40704	234.6	$(128^{10}\, 512^{9}\, 2048^{17})$	(0 7 31)
41216	237.2	$(128^{10}\, 512^{10}\, 2048^{17})$	(0 7 35)
42752	237.9	$(128^{10}\, 512^{9}\, 2048^{18})$	(0 7 31)
43264	240.5	$(128^{10}\, 512^{10}\, 2048^{18})$	(0 7 35)
44800	240.7	$(128^{10}\, 512^{21}\, 4096^{8})$	(0 7 63)
45312	243.4	$(128^{10}\, 512^{22}\, 4096^{8})$	(0 7 67)
48896	244.1	$(128^{10}\, 512^{21}\, 4096^{9})$	(0 7 63)
49408	246.8	$(128^{10}\, 512^{22}\, 4096^{9})$	(0 7 67)
52992	247.5	$(128^{10}\, 512^{21}\, 4096^{10})$	(0 7 63)
53504	250.1	$(128^{10}\, 512^{22}\, 4096^{10})$	(0 7 67)
57088	250.9	$(128^{10}\, 512^{21}\, 4096^{11})$	(0 7 63)
57600	253.5	$(128^{10}\, 512^{22}\, 4096^{11})$	(0 7 67)
61184	254.3	$(128^{10}\, 512^{21}\, 4096^{12})$	(0 7 63)
61696	256.9	$(128^{10}\, 512^{22}\, 4096^{12})$	(0 7 67)
65280	257.7	$(128^{10}\, 512^{21}\, 4096^{13})$	(0 7 63)
65792	260.3	$(128^{10}\, 512^{22}\, 4096^{13})$	(0 7 67)
69376	261.0	$(128^{10}\, 512^{21}\, 4096^{14})$	(0 7 63)
69888	263.7	$(128^{10}\, 512^{22}\, 4096^{14})$	(0 7 67)
73472	264.4	$(128^{10}\, 512^{21}\, 4096^{15})$	(0 7 63)
73984	267.0	$(128^{10}\, 512^{22}\, 4096^{15})$	(0 7 67)
77568	267.8	$(128^{10}\, 512^{21}\, 4096^{16})$	(0 7 63)
78080	270.4	$(128^{10}\, 512^{22}\, 4096^{16})$	(0 7 67)

81664	271.2	$(128^{10}\,512^{21}\,4096^{17})$	$(0\ 7\ 63)$
82176	273.8	$(128^{10}\,512^{22}\,4096^{17})$	$(0\ 7\ 67)$
85760	274.6	$(128^{10}\,512^{21}\,4096^{18})$	$(0\ 7\ 63)$
89856	276.2	$(128^{22}\,1024^{21}\,8192^{8})$	$(0\ 15\ 127)$
90880	278.8	$(128^{22}\,1024^{22}\,8192^{8})$	$(0\ 15\ 135)$
98048	279.6	$(128^{22}\,1024^{21}\,8192^{9})$	$(0\ 15\ 127)$
99072	282.2	$(128^{22}\,1024^{22}\,8192^{9})$	$(0\ 15\ 135)$
106240	283.0	$(128^{22}\,1024^{21}\,8192^{10})$	$(0\ 15\ 127)$
107264	285.6	$(128^{22}\,1024^{22}\,8192^{10})$	$(0\ 15\ 135)$
114432	286.4	$(128^{22}\,1024^{21}\,8192^{11})$	$(0\ 15\ 127)$
115456	288.9	$(128^{22}\,1024^{22}\,8192^{11})$	$(0\ 15\ 135)$
122624	289.8	$(128^{22}\,1024^{21}\,8192^{12})$	$(0\ 15\ 127)$
123648	292.3	$(128^{22}\,1024^{22}\,8192^{12})$	$(0\ 15\ 135)$
130816	293.2	$(128^{22}\,1024^{21}\,8192^{13})$	$(0\ 15\ 127)$
131840	295.7	$(128^{22}\,1024^{22}\,8192^{13})$	$(0\ 15\ 135)$
139008	296.5	$(128^{22}\,1024^{21}\,8192^{14})$	$(0\ 15\ 127)$
140032	299.1	$(128^{22}\,1024^{22}\,8192^{14})$	$(0\ 15\ 135)$
147200	299.9	$(128^{22}\,1024^{21}\,8192^{15})$	$(0\ 15\ 127)$
148224	302.5	$(128^{22}\,1024^{22}\,8192^{15})$	$(0\ 15\ 135)$
155392	303.3	$(128^{22}\,1024^{21}\,8192^{16})$	$(0\ 15\ 127)$
156416	305.9	$(128^{22}\,1024^{22}\,8192^{16})$	$(0\ 15\ 135)$
163584	306.7	$(128^{22}\,1024^{21}\,8192^{17})$	$(0\ 15\ 127)$
164608	309.2	$(128^{22}\,1024^{22}\,8192^{17})$	$(0\ 15\ 135)$
171776	310.1	$(128^{22}\,1024^{21}\,8192^{18})$	$(0\ 15\ 127)$
172800	312.6	$(128^{22}\,1024^{22}\,8192^{18})$	$(0\ 15\ 135)$
179968	313.5	$(128^{22}\,1024^{21}\,8192^{19})$	$(0\ 15\ 127)$
180992	316.0	$(128^{22}\,1024^{22}\,8192^{19})$	$(0\ 15\ 135)$
188160	316.8	$(128^{22}\,1024^{21}\,8192^{20})$	$(0\ 15\ 127)$
189184	319.4	$(128^{22}\,1024^{22}\,8192^{20})$	$(0\ 15\ 135)$
196352	320.2	$(128^{22}\,1024^{21}\,8192^{21})$	$(0\ 15\ 127)$
197376	322.8	$(128^{22}\,1024^{22}\,8192^{21})$	$(0\ 15\ 135)$
204544	323.6	$(128^{22}\,1024^{21}\,8192^{22})$	$(0\ 15\ 127)$
205568	326.2	$(128^{22}\,1024^{22}\,8192^{22})$	$(0\ 15\ 135)$
212736	327.0	$(128^{22}\,1024^{21}\,8192^{23})$	$(0\ 15\ 127)$
213760	329.5	$(128^{22}\,1024^{22}\,8192^{23})$	$(0\ 15\ 135)$
220928	330.4	$(128^{22}\,1024^{21}\,8192^{24})$	$(0\ 15\ 127)$
221952	332.9	$(128^{22}\,1024^{22}\,8192^{24})$	$(0\ 15\ 135)$
229120	333.8	$(128^{22}\,1024^{21}\,8192^{25})$	$(0\ 15\ 127)$
230144	336.3	$(128^{22}\,1024^{22}\,8192^{25})$	$(0\ 15\ 135)$
237312	337.1	$(128^{22}\,1024^{21}\,8192^{26})$	$(0\ 15\ 127)$
238336	339.7	$(128^{22}\,1024^{22}\,8192^{26})$	$(0\ 15\ 135)$
245504	340.5	$(128^{22}\,1024^{21}\,8192^{27})$	$(0\ 15\ 127)$
246528	343.1	$(128^{22}\,1024^{22}\,8192^{27})$	$(0\ 15\ 135)$
253696	343.9	$(128^{22}\,1024^{21}\,8192^{28})$	$(0\ 15\ 127)$
254720	346.5	$(128^{22}\,1024^{22}\,8192^{28})$	$(0\ 15\ 135)$
261888	347.3	$(128^{22}\,1024^{21}\,8192^{29})$	$(0\ 15\ 127)$
262912	349.9	$(128^{22}\,1024^{22}\,8192^{29})$	$(0\ 15\ 135)$
270080	350.7	$(128^{22}\,1024^{21}\,8192^{30})$	$(0\ 15\ 127)$
278272	351.8	$(128^{10}\,512^{21}\,4096^{9}\,16384^{14})$	$(0\ 7\ 63\ 255)$
278784	354.4	$(128^{10}\,512^{22}\,4096^{9}\,16384^{14})$	$(0\ 7\ 67\ 259)$
282368	355.2	$(128^{10}\,512^{21}\,4096^{10}\,16384^{14})$	$(0\ 7\ 63\ 287)$

294656	355.5	$(128^{10}\, 512^{21}\, 4096^9\, 16384^{15})$	(0 7 63 255)
295168	358.2	$(128^{10}\, 512^{22}\, 4096^9\, 16384^{15})$	(0 7 67 259)
298752	358.9	$(128^{10}\, 512^{21}\, 4096^{10}\, 16384^{15})$	(0 7 63 287)
311040	359.2	$(128^{10}\, 512^{21}\, 4096^9\, 16384^{16})$	(0 7 63 255)
311552	361.9	$(128^{10}\, 512^{22}\, 4096^9\, 16384^{16})$	(0 7 67 259)
315136	362.6	$(128^{10}\, 512^{21}\, 4096^{10}\, 16384^{16})$	(0 7 63 287)
327424	363.0	$(128^{10}\, 512^{21}\, 4096^9\, 16384^{17})$	(0 7 63 255)
327936	365.6	$(128^{10}\, 512^{22}\, 4096^9\, 16384^{17})$	(0 7 67 259)
331520	366.3	$(128^{10}\, 512^{21}\, 4096^{10}\, 16384^{17})$	(0 7 63 287)
343808	366.7	$(128^{10}\, 512^{21}\, 4096^9\, 16384^{18})$	(0 7 63 255)
360192	368.2	$(128^{22}\, 1024^{21}\, 8192^9\, 32768^8)$	(0 15 127 511)
361216	370.8	$(128^{22}\, 1024^{22}\, 8192^9\, 32768^8)$	(0 15 135 519)
368384	371.6	$(128^{22}\, 1024^{21}\, 8192^{10}\, 32768^8)$	(0 15 127 575)
392960	372.0	$(128^{22}\, 1024^{21}\, 8192^9\, 32768^9)$	(0 15 127 511)
393984	374.5	$(128^{22}\, 1024^{22}\, 8192^9\, 32768^9)$	(0 15 135 519)
401152	375.3	$(128^{22}\, 1024^{21}\, 8192^{10}\, 32768^9)$	(0 15 127 575)
425728	375.7	$(128^{22}\, 1024^{21}\, 8192^9\, 32768^{10})$	(0 15 127 511)
426752	378.2	$(128^{22}\, 1024^{22}\, 8192^9\, 32768^{10})$	(0 15 135 519)
433920	379.0	$(128^{22}\, 1024^{21}\, 8192^{10}\, 32768^{10})$	(0 15 127 575)
458496	379.4	$(128^{22}\, 1024^{21}\, 8192^9\, 32768^{11})$	(0 15 127 511)
459520	381.9	$(128^{22}\, 1024^{22}\, 8192^9\, 32768^{11})$	(0 15 135 519)
466688	382.8	$(128^{22}\, 1024^{21}\, 8192^{10}\, 32768^{11})$	(0 15 127 575)
491264	383.1	$(128^{22}\, 1024^{21}\, 8192^9\, 32768^{12})$	(0 15 127 511)
492288	385.6	$(128^{22}\, 1024^{22}\, 8192^9\, 32768^{12})$	(0 15 135 519)
499456	386.5	$(128^{22}\, 1024^{21}\, 8192^{10}\, 32768^{12})$	(0 15 127 575)
524032	386.8	$(128^{22}\, 1024^{21}\, 8192^9\, 32768^{13})$	(0 15 127 511)

A.2.5. Practical partitions of variant 2 for the block length B=128

Filter length N	Cost/sample	Optimal partition $\mathcal{P}_{\mathrm{opt}}$	Segment clearances
128	43.0	(128^1)	(0)
256	45.8	(128^2)	(0)
384	48.7	(128^3)	(0)
512	51.6	(128^4)	(0)
768	88.5	$(128^2\, 256^2)$	(0 1)
1024	91.2	$(128^2\, 256^3)$	(0 1)
1280	93.9	$(128^2\, 256^4)$	(0 1)
1536	96.6	$(128^2\, 256^5)$	(0 1)
1792	99.3	$(128^2\, 256^6)$	(0 1)
2048	101.9	$(128^2\, 256^7)$	(0 1)
2304	104.6	$(128^2\, 256^8)$	(0 1)
2560	107.3	$(128^2\, 256^9)$	(0 1)
2816	110.0	$(128^2\, 256^{10})$	(0 1)
3072	112.7	$(128^2\, 256^{11})$	(0 1)
3328	115.4	$(128^2\, 256^{12})$	(0 1)
3584	118.1	$(128^2\, 256^{13})$	(0 1)
3840	120.7	$(128^2\, 256^{14})$	(0 1)

4096	123.4	$(128^2\, 256^{15})$	$(0\ 1)$
4352	126.1	$(128^2\, 256^{16})$	$(0\ 1)$
4608	128.8	$(128^2\, 256^{17})$	$(0\ 1)$
4864	131.5	$(128^2\, 256^{18})$	$(0\ 1)$
5120	134.2	$(128^2\, 256^{19})$	$(0\ 1)$
5376	136.9	$(128^2\, 256^{20})$	$(0\ 1)$
5632	139.5	$(128^2\, 256^{21})$	$(0\ 1)$
5888	142.2	$(128^2\, 256^{22})$	$(0\ 1)$
6144	144.9	$(128^2\, 256^{23})$	$(0\ 1)$
6400	147.3	$(128^2\, 256^4\, 1024^5)$	$(0\ 1\ 3)$
7424	149.8	$(128^2\, 256^4\, 1024^6)$	$(0\ 1\ 3)$
8448	152.4	$(128^2\, 256^4\, 1024^7)$	$(0\ 1\ 3)$
9472	154.9	$(128^2\, 256^4\, 1024^8)$	$(0\ 1\ 3)$
10496	157.5	$(128^2\, 256^4\, 1024^9)$	$(0\ 1\ 3)$
11520	160.1	$(128^2\, 256^4\, 1024^{10})$	$(0\ 1\ 3)$
12544	162.6	$(128^2\, 256^4\, 1024^{11})$	$(0\ 1\ 3)$
13568	165.2	$(128^2\, 256^4\, 1024^{12})$	$(0\ 1\ 3)$
14592	167.7	$(128^2\, 256^4\, 1024^{13})$	$(0\ 1\ 3)$
15616	170.3	$(128^2\, 256^4\, 1024^{14})$	$(0\ 1\ 3)$
16640	172.9	$(128^2\, 256^4\, 1024^{15})$	$(0\ 1\ 3)$
17664	175.4	$(128^2\, 256^4\, 1024^{16})$	$(0\ 1\ 3)$
18688	178.0	$(128^2\, 256^4\, 1024^{17})$	$(0\ 1\ 3)$
19712	180.5	$(128^2\, 256^4\, 1024^{18})$	$(0\ 1\ 3)$
20736	181.7	$(128^2\, 256^8\, 2048^9)$	$(0\ 1\ 3)$
20992	184.4	$(128^2\, 256^9\, 2048^9)$	$(0\ 1\ 5)$
22784	185.1	$(128^2\, 256^8\, 2048^{10})$	$(0\ 1\ 3)$
23040	187.8	$(128^2\, 256^9\, 2048^{10})$	$(0\ 1\ 5)$
24832	188.4	$(128^2\, 256^8\, 2048^{11})$	$(0\ 1\ 3)$
25088	191.1	$(128^2\, 256^9\, 2048^{11})$	$(0\ 1\ 5)$
26880	191.8	$(128^2\, 256^8\, 2048^{12})$	$(0\ 1\ 3)$
27136	194.5	$(128^2\, 256^9\, 2048^{12})$	$(0\ 1\ 5)$
28928	195.1	$(128^2\, 256^8\, 2048^{13})$	$(0\ 1\ 3)$
29184	197.8	$(128^2\, 256^9\, 2048^{13})$	$(0\ 1\ 5)$
30976	198.5	$(128^2\, 256^8\, 2048^{14})$	$(0\ 1\ 3)$
31232	201.2	$(128^2\, 256^9\, 2048^{14})$	$(0\ 1\ 5)$
33024	201.3	$(128^2\, 256^{16}\, 4096^7)$	$(0\ 1\ 3)$
33280	204.0	$(128^2\, 256^{17}\, 4096^7)$	$(0\ 1\ 5)$
37120	204.7	$(128^2\, 256^{16}\, 4096^8)$	$(0\ 1\ 3)$
37376	207.4	$(128^2\, 256^{17}\, 4096^8)$	$(0\ 1\ 5)$
41216	208.1	$(128^2\, 256^{16}\, 4096^9)$	$(0\ 1\ 3)$
41472	210.8	$(128^2\, 256^{17}\, 4096^9)$	$(0\ 1\ 5)$
45312	211.5	$(128^2\, 256^{16}\, 4096^{10})$	$(0\ 1\ 3)$
45568	214.2	$(128^2\, 256^{17}\, 4096^{10})$	$(0\ 1\ 5)$
49408	214.9	$(128^2\, 256^{16}\, 4096^{11})$	$(0\ 1\ 3)$
49664	217.6	$(128^2\, 256^{17}\, 4096^{11})$	$(0\ 1\ 5)$
53504	218.3	$(128^2\, 256^{16}\, 4096^{12})$	$(0\ 1\ 3)$
53760	220.9	$(128^2\, 256^{17}\, 4096^{12})$	$(0\ 1\ 5)$
57600	221.6	$(128^2\, 256^{16}\, 4096^{13})$	$(0\ 1\ 3)$
57856	224.3	$(128^2\, 256^{17}\, 4096^{13})$	$(0\ 1\ 5)$
61696	225.0	$(128^2\, 256^{16}\, 4096^{14})$	$(0\ 1\ 3)$
61952	227.7	$(128^2\, 256^{17}\, 4096^{14})$	$(0\ 1\ 5)$

65792	228.4	$(128^2\ 256^{16}\ 4096^{15})$	(0 1 3)
66816	230.3	$(128^2\ 256^4\ 1024^8\ 8192^7)$	(0 1 3 11)
69888	231.8	$(128^2\ 256^{16}\ 4096^{16})$	(0 1 3)
75008	233.6	$(128^2\ 256^4\ 1024^8\ 8192^8)$	(0 1 3 11)
76032	236.2	$(128^2\ 256^4\ 1024^9\ 8192^8)$	(0 1 3 19)
83200	237.0	$(128^2\ 256^4\ 1024^8\ 8192^9)$	(0 1 3 11)
84224	239.6	$(128^2\ 256^4\ 1024^9\ 8192^9)$	(0 1 3 19)
91392	240.4	$(128^2\ 256^4\ 1024^8\ 8192^{10})$	(0 1 3 11)
92416	243.0	$(128^2\ 256^4\ 1024^9\ 8192^{10})$	(0 1 3 19)
99584	243.8	$(128^2\ 256^4\ 1024^8\ 8192^{11})$	(0 1 3 11)
100608	246.4	$(128^2\ 256^4\ 1024^9\ 8192^{11})$	(0 1 3 19)
107776	247.2	$(128^2\ 256^4\ 1024^8\ 8192^{12})$	(0 1 3 11)
108800	249.7	$(128^2\ 256^4\ 1024^9\ 8192^{12})$	(0 1 3 19)
115968	250.6	$(128^2\ 256^4\ 1024^8\ 8192^{13})$	(0 1 3 11)
116992	253.1	$(128^2\ 256^4\ 1024^9\ 8192^{13})$	(0 1 3 19)
124160	253.9	$(128^2\ 256^4\ 1024^8\ 8192^{14})$	(0 1 3 11)
125184	256.5	$(128^2\ 256^4\ 1024^9\ 8192^{14})$	(0 1 3 19)
132352	257.1	$(128^2\ 256^4\ 1024^{16}\ 16384^7)$	(0 1 3 11)
133376	259.7	$(128^2\ 256^4\ 1024^{17}\ 16384^7)$	(0 1 3 19)
140544	260.7	$(128^2\ 256^4\ 1024^8\ 8192^{16})$	(0 1 3 11)
148736	260.9	$(128^2\ 256^4\ 1024^{16}\ 16384^8)$	(0 1 3 11)
149760	263.4	$(128^2\ 256^4\ 1024^{17}\ 16384^8)$	(0 1 3 19)
165120	264.6	$(128^2\ 256^4\ 1024^{16}\ 16384^9)$	(0 1 3 11)
166144	267.1	$(128^2\ 256^4\ 1024^{17}\ 16384^9)$	(0 1 3 19)
181504	268.3	$(128^2\ 256^4\ 1024^{16}\ 16384^{10})$	(0 1 3 11)
182528	270.8	$(128^2\ 256^4\ 1024^{17}\ 16384^{10})$	(0 1 3 19)
197888	272.0	$(128^2\ 256^4\ 1024^{16}\ 16384^{11})$	(0 1 3 11)
198912	274.5	$(128^2\ 256^4\ 1024^{17}\ 16384^{11})$	(0 1 3 19)
214272	275.7	$(128^2\ 256^4\ 1024^{16}\ 16384^{12})$	(0 1 3 11)
215296	278.3	$(128^2\ 256^4\ 1024^{17}\ 16384^{12})$	(0 1 3 19)
230656	279.4	$(128^2\ 256^4\ 1024^{16}\ 16384^{13})$	(0 1 3 11)
231680	282.0	$(128^2\ 256^4\ 1024^{17}\ 16384^{13})$	(0 1 3 19)
247040	283.1	$(128^2\ 256^4\ 1024^{16}\ 16384^{14})$	(0 1 3 11)
248064	285.7	$(128^2\ 256^4\ 1024^{17}\ 16384^{14})$	(0 1 3 19)
263424	286.8	$(128^2\ 256^4\ 1024^{16}\ 16384^{15})$	(0 1 3 11)
264448	289.4	$(128^2\ 256^4\ 1024^{17}\ 16384^{15})$	(0 1 3 19)
266496	289.7	$(128^2\ 256^{16}\ 4096^8\ 32768^7)$	(0 1 3 35)
279808	290.5	$(128^2\ 256^4\ 1024^{16}\ 16384^{16})$	(0 1 3 11)
280832	293.1	$(128^2\ 256^4\ 1024^{17}\ 16384^{16})$	(0 1 3 19)
299264	293.4	$(128^2\ 256^{16}\ 4096^8\ 32768^8)$	(0 1 3 35)
299520	296.0	$(128^2\ 256^{17}\ 4096^8\ 32768^8)$	(0 1 5 37)
303360	296.7	$(128^2\ 256^{16}\ 4096^9\ 32768^8)$	(0 1 3 67)
332032	297.1	$(128^2\ 256^{16}\ 4096^8\ 32768^9)$	(0 1 3 35)
332288	299.8	$(128^2\ 256^{17}\ 4096^8\ 32768^9)$	(0 1 5 37)
336128	300.4	$(128^2\ 256^{16}\ 4096^9\ 32768^9)$	(0 1 3 67)
364800	300.8	$(128^2\ 256^{16}\ 4096^8\ 32768^{10})$	(0 1 3 35)
365056	303.5	$(128^2\ 256^{17}\ 4096^8\ 32768^{10})$	(0 1 5 37)
368896	304.2	$(128^2\ 256^{16}\ 4096^9\ 32768^{10})$	(0 1 3 67)
397568	304.5	$(128^2\ 256^{16}\ 4096^8\ 32768^{11})$	(0 1 3 35)
397824	307.2	$(128^2\ 256^{17}\ 4096^8\ 32768^{11})$	(0 1 5 37)
401664	307.9	$(128^2\ 256^{16}\ 4096^9\ 32768^{11})$	(0 1 3 67)

430336	308.2	$(\,128^2\,256^{16}\,4096^8\,32768^{12}\,)$	$(\,0\ 1\ 3\ 35\,)$
430592	310.9	$(\,128^2\,256^{17}\,4096^8\,32768^{12}\,)$	$(\,0\ 1\ 5\ 37\,)$
434432	311.6	$(\,128^2\,256^{16}\,4096^9\,32768^{12}\,)$	$(\,0\ 1\ 3\ 67\,)$
463104	311.9	$(\,128^2\,256^{16}\,4096^8\,32768^{13}\,)$	$(\,0\ 1\ 3\ 35\,)$
463360	314.6	$(\,128^2\,256^{17}\,4096^8\,32768^{13}\,)$	$(\,0\ 1\ 5\ 37\,)$
467200	315.3	$(\,128^2\,256^{16}\,4096^9\,32768^{13}\,)$	$(\,0\ 1\ 3\ 67\,)$
495872	315.6	$(\,128^2\,256^{16}\,4096^8\,32768^{14}\,)$	$(\,0\ 1\ 3\ 35\,)$
496128	318.3	$(\,128^2\,256^{17}\,4096^8\,32768^{14}\,)$	$(\,0\ 1\ 5\ 37\,)$
499968	319.0	$(\,128^2\,256^{16}\,4096^9\,32768^{14}\,)$	$(\,0\ 1\ 3\ 67\,)$
528640	319.3	$(\,128^2\,256^{16}\,4096^8\,32768^{15}\,)$	$(\,0\ 1\ 3\ 35\,)$

A.3. Identities

Any sequence $x(n)$ can be decomposed into an even part $x_{\text{even}}(n)$ and an odd part $x_{\text{odd}}(n)$ using the well-known relation [74]

$$x(n) = x_{\text{even}}(n) + x_{\text{odd}}(n) \quad \text{with} \quad \begin{aligned} x_{\text{even}}(n) &= \frac{x(n) + x(N - n)}{2} \\[2mm] x_{\text{odd}}(n) &= \frac{x(n) - x(N - n)}{2} \end{aligned} \tag{A.3}$$

Let $N, i, j \in \mathbb{Z}$ $(i, j < N)$ be integers. The following equivalences hold for series of powers-of-two and of multiples of powers-of-two:

$$\sum_{i=0}^{N-1} 2^i = 2^N - 1 \qquad \text{(Geometric series)} \tag{A.4}$$

$$\sum_{i=j}^{N-1} 2^i = \sum_{i=0}^{N-1} 2^i - \sum_{i=0}^{j-1} 2^i = 2^N - 2^j \tag{A.5}$$

$$\sum_{i=0}^{N-1} i2^i = \sum_{j=1}^{N-1}\sum_{i=j}^{N-1} 2^i = \sum_{j=1}^{N-1} \left(2^N - 2^j \right) = \sum_{j=1}^{N-1} 2^N - \sum_{j=1}^{N-1} 2^j$$

$$= (N - 1)2^N - (2^N - 2) = N2^N - 2^{N+1} + 2 \tag{A.6}$$

Bibliography

[1] Intel Integrated Performance Primitives Reference Manual: Volume 1: Signal Processing. http://software.intel.com/sites/products/documentation/doclib/ipp_sa/80/ipps.pdf, 2013. [online, last accessed on April 17, 2015].

[2] F. Adriaensen. Design of a convolution engine optimised for reverb. In 4th International Linux Audio Conference (LAC2006), Karlsruhe, Germany, 2006.

[3] R. C. Agarwal and C. S. Burrus. Fast convolution using Fermat number transforms with applications to digital filtering. IEEE Transactions on Acoustics, Speech and Signal Processing, Vol. 22(2):87–97, 1974.

[4] R. C. Agarwal and C. S. Burrus. Number theoretic transforms to implement fast digital convolution. Proceedings of the IEEE, Vol. 63(4):550–560, 1975.

[5] R. C. Agarwal and J. W. Cooley. New algorithms for digital convolution. IEEE Transactions on Acoustics, Speech and Signal Processing, Vol. 25(5):392–410, 1977.

[6] J. B. Allen and D. A. Berkley. Image method for efficiently simulating small-room acoustics. Journal of the Acoustical Society of America (JASA), Vol. 65(4):943–950, 1979.

[7] G. E. Andrews. The Theory of Partitions. Cambridge Mathematical Library. Cambridge University Press, 1998.

[8] E. Armelloni, C. Giottoli, and A. Farina. Implementation of real-time partitioned convolution on a DSP board. In IEEE Workshop on Applications of Signal Processing to Audio and Acoustics (WASPAA'03), New Paltz, USA, 2003.

[9] E. Battenberg and R. Avizienis. Implementing Real-Time Partitioned Convolution Algorithms on Conventional Operating Systems. In 14th International Conference on Digital Audio Effects (DAFx-11), Paris, France, 2011.

[10] C. D. Bergland. A Fast Fourier Transform Algorithm Using Base 8 Iterations. Mathematics of Computation, Vol. 22:275–275, 1968.

[11] M. Bhattacharya and R. C. Agarwal. Number theoretic techniques for

computation of digital convolution. *IEEE Transactions on Acoustics, Speech and Signal Processing*, Vol. 32(3):507–511, 1984.

[12] M. Bhattacharya, R. Creutzburg, and J. Astola. Some historical notes on number theoretic transform. In *Proc. 2004 Int. TICS Workshop on Spectral Methods and Multirate Signal Processing*, 2004.

[13] R. E. Blahut. *Fast algorithms for digital signal processing*. Addison-Wesley, 1st edition, 1985.

[14] L. I. Bluestein. A linear filtering approach to the computation of discrete Fourier transform. *IEEE Transactions on Audio and Electroacoustics*, Vol. 18(4):451–455, 1970.

[15] D. Botteldooren. Acoustical finite-difference time-domain simulation in a quasi-cartesian grid. *Journal of the Acoustical Society of America (JASA)*, Vol. 95(5):2313–2319, 1994.

[16] R. N. Bracewell. Discrete Hartley transform. *Journal of the Optical Society of America*, Volume 73:1832–1835, 1983.

[17] R. N. Bracewell. The fast Hartley transform. *Proceedings of the IEEE*, Vol. 72(8):1010–1018, 1984.

[18] V. Britanak, P. Yip, and K. R. Rao. *Discrete Cosine and Sine Transforms: General Properties, Fast Algorithms and Integer Approximations*. Academic Press Professional, San Diego, USA, 2007.

[19] C. S. Burrus. Index mappings for multidimensional formulation of the DFT and convolution. *IEEE Transactions on Acoustics, Speech and Signal Processing*, Vol. 25(3):239–242, 1977.

[20] C. S. Burrus. Index mappings for multidimensional formulation of the dft and convolution. *IEEE Transactions on Acoustics, Speech and Signal Processing*, Vol. 25(3), 1977.

[21] C. S. Burrus and T. W. Parks. *DFT/FFT and convolution algorithms: Theory and Implementation*. John Wiley & Sons, 1984.

[22] S. S. Chandra. Finite Transform Library (FTL) homepage. `http://finitetransform.sourceforge.net`. [online, last accessed on April 17, 2015].

[23] S. A. Cook. *On the minimum computation time of functions*. PhD thesis, Department of Mathematics, Harvard University, 1966.

[24] J. W. Cooley and J. W. Tukey. An Algorithm for the Machine Calculation of Complex Fourier Series. *Mathematics of Computation*, Vol. 19(90), 1965.

[25] T. H. Cormen, C. Stein, R. L. Rivest, and C. E. Leiserson. *Introduction to Algorithms*. McGraw-Hill Higher Education, 2nd edition, 2001.

[26] J. Dongarra and F. Sullivan. Top Ten Algorithms of the Century. *Computing in Science and Engineering*, Vol. 2(1), 2000.

[27] P. Duhamel and H. Hollmann. 'Split radix' FFT algorithm. *Electronics Letters*, Vol. 20(1):14–16, 1984.

[28] P. Duhamel and M. Vetterli. Improved Fourier and Hartley transform algorithms: Application to cyclic convolution of real data. *IEEE Transactions on Acoustics, Speech and Signal Processing*, Vol. 35(6), 1987.

[29] P. Duhamel and M. Vetterli. Fast Fourier Transforms: A Tutorial Review and a State of the Art. *Signal Processing*, Vol. 19(4):259–299, 1990.

[30] G. P. M. Egelmeers and P. C. W. Sommen. A new method for efficient convolution in frequency domain by non-uniform partitioning. In *proceedings of the 7th European Signal Processing Conference*, pages 1030–1033, Edinburgh, United Kingdom, 1994.

[31] G. P. M. Egelmeers and P. C. W. Sommen. A new method for efficient convolution in frequency domain by nonuniform partitioning for adaptive filtering. *IEEE Transactions on Signal Processing*, 44(12):3123–3129, 1996.

[32] S. H. Foster and E. M. Wenzel. The Convolvotron: Real-time demonstration of reverberant virtual acoustic environments. *Journal of the Acoustical Society of America (JASA)*, Vol. 92(4):2376–2376, 1992.

[33] F. Franchetti and M. Püschel. Generating high performance pruned FFT implementations. In *IEEE International Conference on Acoustics, Speech and Signal Processing (ICASSP 2009)*, Taipeh, Taiwan, 2009.

[34] M. Frigo and S. G. Johnson. FFTW3 homepage. http://www.fftw.org. [online, last accessed on April 17, 2015].

[35] M. Frigo and S. G. Johnson. Pruned FFTs, FFTW3 homepage. http://www.fftw.org/pruned.html. [online, last accessed on April 17, 2015].

[36] M. Frigo and S. G. Johnson. The Design and Implementation of FFTW3. *Proceedings of the IEEE*, Vol. 93(2):216–231, 2005.

[37] G. García. Optimal filter partition for efficient convolution with short input/output delay. In *Audio Engineering Society Convention 113*, Los Angeles, USA, 2002.

[38] W. G. Gardner. Efficient convolution without input/output delay. In *Audio Engineering Society Convention 97*, San Francisco, USA, 1994.

[39] W. G. Gardner. Efficient convolution without input-output delay. *Jour-*

nal of the Audio Engineering Society, Vol. 43:127–136, 1995.

[40] W. M. Gentleman and G. Sande. Fast Fourier Transforms: For Fun and Profit. In *1966 Fall Joint Computer Conference*, San Francisco, USA, 1966.

[41] G. Goertzel. An algorithm for the evaluation of finite trigonometric series. *American Mathematical Monthly*, Vol. 65(I):34–35, 1958.

[42] I. J. Good. The interaction algorithm and practical Fourier analysis. *Journal of the Royal Statistical Society*, Vol. 20:361–372, 1958.

[43] R. M. Gray. Toeplitz and circulant matrices: A review. *Communications and Information Theory*, Vol. 2(3):155–239, 2005.

[44] S. Gudvangen. Number Theoretic Transforms in Audio Processing. In 2nd *International Conference on Digital Audio Effects (DAFx-99)*, Trondheim, Norway, 1999.

[45] M. Heideman, D. H. Johnson, and C. S. Burrus. Gauss and the History of the Fast Fourier Transform. *IEEE ASSP Magazine*, Vol. 1(4):14–21, 1984.

[46] J. Huopaniemi, L. Savioja, and T. Takala. DIVA Virtual Audio Reality System. In *International Conference on Auditory Display (ICAD'96)*, Palo Alto, USA, 1996.

[47] J. Hurchalla. Low latency convolution in one dimension via two dimensional convolutions: An intuitive approach. In *Audio Engineering Society Convention 125*, San Francisco, USA, 2008.

[48] T. Ilmonen and T. Lokki. Extreme Filters—cache-efficient implementation of long IIR and FIR filters. *IEEE Signal Processing Letters*, Vol. 13(7), 2006.

[49] I. Ito and H. Kiya. A computing method for linear convolution in the DCT domain. In 19th *European Signal Processing Conference (EUSIPCO 2011)*, Barcelona, Spain, 2011.

[50] S. G. Johnson and M. Frigo. A Modified Split-Radix FFT With Fewer Arithmetic Operations. *IEEE Transactions on Signal Processing*, 55(1):111–119, 2007.

[51] M. Joho and G. S. Moschytz. Connecting partitioned frequency-domain filters in parallel or in cascade. *IEEE Transactions on Circuits and Systems II: Analog and Digital Signal Processing*, Vol. 47(8):685–698, 2000.

[52] J.-M. Jot and A. Chaigne. Digital delay networks for designing artificial reverberators. In *Audio Engineering Society Convention 90*, Paris, France, 1991.

[53] J.-M. Jot, V. Larcher, and O. Warusfel. Digital signal processing issues in the context of binaural and transaural stereophony. In *Audio Engineering Society Convention 98*, Paris, France, 1995.

[54] A. Karatsuba and Y. Ofman. Multiplication of multidigit numbers on automata. *Soviet Physics-Doklady*, Vol. 7:595–596, 1963.

[55] M. Karjalainen, T. Lokki, H. Nironen, A. Harma, L. Savioja, and S. Vesa. Application Scenarios of Wearable and Mobile Augmented Reality Audio. In *Audio Engineering Society Convention 116*, Berlin, Germany, 2004.

[56] M. Kleiner, B.-I. Dalenbäck, and P. Svensson. Auralization—An Overview. *Journal of the Audio Engineering Society*, Vol. 41(11):861–875, 1993.

[57] D. E. Knuth. *The Art of Computer Programming, Volume 2: Seminumerical Algorithms*. Addison-Wesley Longman Publishing, Boston, USA, 3rd edition, 1997.

[58] A. Krokstad, S. Strom, and S. Sørsdal. Calculating the acoustical room response by the use of a ray tracing technique. *Journal of Sound and Vibration*, Vol. 8(1):118–125, 1968.

[59] B. D. Kulp. Digital equalization using fourier transform techniques. In *Audio Engineering Society Convention 85*, Los Angeles, USA, 1988.

[60] H. Kuttruff. *Room acoustics*. CRC Press, 2009.

[61] W.-C. Lee, C.-M. Liu, C.-H. Yang, and J.-I. Guo. Fast perceptual convolution for room reverberation. In 6th *International Conference on Digital Audio Effects (DAFx-03)*, London, United Kingdom, 2003.

[62] T. Lentz and G. Behler. Dynamic cross-talk cancellation for binaural synthesis in virtual reality environments. In *Audio Engineering Society Convention 117*, San Francisco, USA, 2004.

[63] T. Lentz, D. Schröder, G. Behler, and M. Vorländer. Real-time audio rendering system for virtual reality. *Journal of the Acoustical Society of America (JASA)*, Vol. 120(5), 2006.

[64] T. Lentz, D. Schröder, M. Vorländer, and I. Assenmacher. Virtual reality system with integrated sound field simulation and reproduction. *EURASIP journal on applied signal processing*, Vol. 2007, 2007.

[65] J. Markel. FFT pruning. *IEEE Transactions on Audio and Electroacoustics*, Vol. 19(4), 1971.

[66] S. A. Martucci. Symmetric convolution and the discrete sine and cosine transforms. *IEEE Transactions on Signal Processing*, Vol. 42(5):1038–1051, 1994.

[67] S. A. Martucci. Digital filtering of images using the discrete sine or cosine transform. *Optical Engineering*, Vol. 35(1):119–127, 1996.

[68] J. H. McClellan and C. M. Rader. *Number theory in digital signal processing*. Prentice-Hall, 1979.

[69] D. S. McGrath. Huron - A Digital Audio Convolution Workstation. In *Audio Engineering Society Regional Convention*, Syndey, Australia, 1995.

[70] Q. Mo, M. Taylor, A. Chandak, C. Lauterbach, C. Schissler, and D. Manocha. Interactive GPU-based sound auralization in dynamic scenes. *Journal of the Acoustical Society of America (JASA)*, Vol. 133(5):3614–3614, 2013.

[71] C. Müller-Tomfelde. Low-latency convolution for real-time applications. In 16th *Audio Engineering Society Conference on Spatial Sound Reproduction*, Rovaniemi, Finland, 1999.

[72] C. Müller-Tomfelde. Time-varying filter in non-uniform block convolution. In 4th *International Conference on Digital Audio Effects (DAFx-01)*, Limerick, Ireland, 2001.

[73] H. J. Nussbaumer. *Fast Fourier transform and convolution algorithms*. Springer series in information sciences. Springer-Verlag, 1981.

[74] A. V. Oppenheim and R. W. Schafer. *Discrete-Time Signal Processing*. Prentice Hall Signal Processing Series. Prentice Hall, 1989.

[75] J. M. Pollard. The Fast Fourier Transform in a Finite Field. *Mathematics of Computation*, Vol. 25(114), 1971.

[76] A. Primavera, S. Cecchi, L. Romoli, P. Peretti, and F. Piazza. A low latency implementation of a non uniform partitioned overlap and save algorithm for real time applications. In *Audio Engineering Society Convention 131*, New York, USA, 2011.

[77] C. M. Rader. Discrete Fourier transforms when the number of data samples is prime. *Proceedings of the IEEE*, Vol. 56(6):1107–1108, 1968.

[78] K. R. Rao and P. Yip. *Discrete Cosine Transform: Algorithms, Advantages, Applications*. Academic Press Professional, San Diego, USA, 1990.

[79] A. J. Reijnen, J.-J. Sonke, and D. de Vries. New developments in electroacoustic reverberation technology. In *Audio Engineering Society Convention 98*, Paris, France, 1995.

[80] A. Reilly and D. McGrath. Convolution Processing for Realistic Reverberation. In *Audio Engineering Society Convention 98*, Paris, France, 1995.

[81] V.G. Reju, Soo N. K., and I.Y. Soon. Convolution using discrete sine and cosine transforms. *IEEE Signal Processing Letters*, Vol. 14(7):445–448, 2007.

[82] M. Sadreddini. Non-Uniformly Partitioned Block Convolution on Graphics Processing Units. Master's thesis, Department of Applied Signal Processing, Blekinge Institute of Technology, 2013.

[83] C. Sander, F. Wefers, and D. Leckschat. Scalable binaural synthesis on mobile devices. In *Audio Engineering Society Convention 133*, San Francisco, USA, 2012.

[84] L. Savioja. Real-time 3D finite-difference time-domain simulation of low- and mid-frequency room acoustics. In 13th *International Conference on Digital Audio Effects (DAFx-10)*, Graz, Austria, 2010.

[85] L. Savioja, J. Huopaniemi, T. Lokki, and R. Väänänen. Creating Interactive Virtual Acoustic Environments. *Journal of the Audio Engineering Society*, Vol. 47(9):675–705, 1999.

[86] L. Savioja, V. Välimäki, and J. O. Smith. Audio signal processing using graphics processing units. *Journal of the Audio Engineering Society*, Vol. 59(1/2):3–19, 2011.

[87] A. Schönhage and V. Strassen. Schnelle Multiplikation großer Zahlen. *Computing*, Vol. 7(3):281–292, 1971.

[88] D. Schröder. *Physically based real-time auralization of interactive virtual environments / Dirk Schröder*. PhD thesis, RWTH Aachen University, 2011.

[89] D. Schröder, F. Wefers, S. Pelzer, D. Rausch, M. Vorländer, and T. Kuhlen. Virtual Reality System at RWTH Aachen University. In 20th *International Congress on Acoustics (ICA2010), Sydney, Australia*, 2010.

[90] D. Schröder, F. Wefers, S. Pelzer, D. Rausch, M. Vorländer, and T. Kuhlen. Virtual reality system at rwth aachen university. In 20th *International Congress on Acoustics (ICA 2010), Sydney, Australia*, 2010.

[91] M. R. Schroeder. Natural sounding artificial reverberation. *Journal of the Audio Engineering Society*, Vol. 10(3):219–223, 1962.

[92] M.R. Schroeder and B. Logan. "colorless" artificial reverberation. *IRE Transactions on Audio*, AU-9(6):209–214, 1961.

[93] R. Sedgewick. *Algorithms*. Addison-Wesley Series in Computer Science. Addision Wesley Publishing Company, 1984.

[94] X. Shao and S. G. Johnson. Type-II/III DCT/DST Algorithms with

Reduced Number of Arithmetic Operations. *Signal Processing*, Vol. 88(6):1553–1564, 2008.

[95] X. Shao and S. G. Johnson. Type-IV DCT, DST, and MDCT Algorithms with Reduced Numbers of Arithmetic Operations. *Signal Processing*, Vol. 88(6):1313–1326, 2008.

[96] D. P. Skinner. Pruning the decimation in-time FFT algorithm. *IEEE Transactions on Acoustics, Speech and Signal Processing*, Vol. 24(2):193–194, 1976.

[97] J. O. Smith. A New Approach to Digital Reverberation Using Closed Waveguide Networks. In *International Computer Music Conference*, 1985.

[98] J.-S. Soo and K.K. Pang. A new structure for block FIR adaptive digital filters. In *IREECON Conference 1987*, Sydney, Australia, 1987.

[99] J.-S. Soo and K.K. Pang. Multidelay block frequency domain adaptive filter. *IEEE Transactions on Acoustics, Speech and Signal Processing*, Vol. 38(2):373–376, 1990.

[100] H. V. Sorensen and C. S. Burrus. Efficient computation of the DFT with only a subset of input or output points. *IEEE Transactions on Signal Processing*, Vol. 41(3):1184–1200, 1993.

[101] H. V. Sorensen, D. L. Jones, C. S. Burrus, and M. Heideman. On computing the discrete Hartley transform. *IEEE Transactions on Acoustics, Speech and Signal Processing*, Vol. 33(5), 1985.

[102] H. V. Sorensen, D. L. Jones, M. Heideman, and C. S. Burrus. Real-valued fast fourier transform algorithms. *IEEE Transactions on Acoustics, Speech and Signal Processing*, Vol. 35(6), 1987.

[103] T. V. Sreenivas and P. Rao. Fft algorithm for both input and output pruning. *IEEE Transactions on Acoustics, Speech and Signal Processing*, Vol. 27(3):291–292, 1979.

[104] E. Stavrakis, N. Tsingos, and P. Calamia. Topological sound propagation with reverberation graphs. *Acta Acustica/Acustica - the Journal of the European Acoustics Association (EAA)*, 2008.

[105] T. G. Stockham Jr. High-speed convolution and correlation. In *Proceedings of the April 26-28, 1966, Spring joint computer conference*, pages 229–233. ACM, 1966.

[106] Holger Strauss. Implementing Doppler Shifts for Virtual Auditory Environments. In *Audio Engineering Society Convention 104*, 1998.

[107] K. Suresh and T. V. Sreenivas. Block convolution using discrete trigonometric transforms and discrete fourier transform. *IEEE Signal*

Processing Letters, Vol. 15:469–472, 2008.

[108] M. Taylor, A. Chandak, L. Antani, and D. Manocha. RESound: Interactive Sound Rendering for Dynamic Virtual Environments. In 17th *ACM International Conference on Multimedia (MM'09)*, pages 271–280, Beijing, China, 2009.

[109] L. H. Thomas. Using a computer to solve problems in physics. In *Applications of Digital Computers*, Boston, USA, 1963.

[110] A. L. Toom. The complexity of a scheme of functional elements realizing the multiplication of integers. *Soviet Mathematics Doklady*, Vol. 3:714–716, 1963.

[111] A. Torger. BruteFIR homepage. `http://www.ludd.luth.se/~torger/brutefir.html`. [online, last accessed on April 17, 2015].

[112] A. Torger and A. Farina. Real-time partitioned convolution for ambiophonics surround sound. In *IEEE Workshop on Applications of Signal Processing to Audio and Acoustics (WASPAA'01)*, New Paltz, USA, 2001.

[113] N. Tsingos. Using Programmable Graphics Hardware for Acoustics and Audio Rendering. In *Audio Engineering Society Convention 127*, New York, USA, 2009.

[114] N. Tsingos, E. Gallo, and G. Drettakis. Perceptual Audio Rendering of Complex Virtual Environments. *ACM Transactions on Graphics (SIGGRAPH Conference Proceedings)*, 23(3), 2004.

[115] M. Vetterli and H. J Nussbaumer. Simple FFT and DCT algorithms with reduced number of operations. *Signal processing*, Vol. 6(4):267–278, 1984.

[116] M. Vorländer. *Auralization: Fundamentals of Acoustics, Modelling, Simulation, Algorithms and Acoustic Virtual Reality*. RWTHedition Series. Springer, 2011.

[117] M. Vorländer. Simulation of the transient and steady-state sound propagation in rooms using a new combined ray-tracing/image-source algorithm. *Journal of the Acoustical Society of America (JASA)*, Vol. 86(1):172–178, 1989.

[118] F. Wefers and J. Berg. High-performance real-time FIR-filtering using fast convolution on graphics hardware. In 13th *International Conference on Digital Audio Effects (DAFx-10)*, Graz, Austria, 2010.

[119] F. Wefers and D. Schröder. Real-time auralization of coupled rooms. In *EAA Auralization Symposium 2009*, Espoo, Finland, 2009.

[120] F. Wefers, J. Stienen, and M. Vorländer. Interactive acoustic virtual

environments using distributed room acoustics simulations. In *EAA Joint Symposium on Auralization and Ambisonics*, Berlin, Germany, 2014.

[121] F. Wefers and M. Vorländer. Optimal filter partitions for non-uniformly partitioned convolution. In *45th Audio Engineering Society Conference on Applications of Time-Frequency Processing in Audio*, Helsinki, Finland, 2012.

[122] F. Wefers and M. Vorländer. Potential of Non-Uniformly Partitioned Convolution with Freely Adaptable FFT Sizes. In *Audio Engineering Society Convention 133*, San Francisco, USA, 2012.

[123] F. Wefers and M. Vorländer. Frequency domain filter exchange for DFT-based fast convolution. In *AIA-DAGA 2013: Conference on Acoustics including the 40th Italian (AIA) Annual Conference on Acoustics and the 39th German Annual Conference on Acoustics (DAGA)*, pages 263–263, Merano, Italy, 2013.

[124] F. Wefers and M. Vorländer. Optimal filter partitions for real-time FIR filtering using uniformly-partitioned FFT-based convolution in the frequency-domain. In *14th International Conference on Digital Audio Effects (DAFx-11)*, Paris, France, 2011.

[125] E. M. Wenzel, J. D. Miller, and J. S. Abel. Sound Lab: A Real-Time, Software-Based System for the Study of Spatial Hearing. In *Audio Engineering Society Convention 108*, Paris, France, 2000.

[126] S. Winograd. Some bilinear forms whose multiplicative complexity depends on the field of constants. *Mathematical Systems Theory*, 10:169–180, 1977.

[127] S. Winograd. On computing the discrete Fourier transform. *Mathematics of computation*, Vol. 32(141):175–199, 1978.

[128] C.-H. Yang, J.-I. Guo, W.-C. Lee, and C.-M. Liu. Perceptual convolution for reverberation. In *Audio Engineering Society Convention 115*, New York, USA, 2003.

[129] R. Yavne. An economical method for calculating the discrete Fourier transform. In *Proceedings of the 1968 fall joint computer conference (AFIPS '68, fall, part I)*, San Francisco, USA, 1968.

[130] U. Zölzer. *Digital Audio Signal Processing*. Wiley and sons, 2nd edition, 2008.

[131] X. Zou, S. Muramatsu, and H. Kiya. The generalized overlap-add and overlap-save methods using discrete sine and cosine transforms for fir filtering. In *1996 3rd International Conference on Signal Processing (ICSP '96)*, volume Vol. 1, pages 91–94, Beijing, China, 1996.

Acknowledgements

This thesis would not have become reality without the help and encouragement of friends and supporters. I like to express my greatest gratitude to all of you, who helped me on my way...

First of all, the biggest thanks go to my wife Nise, who lifted much weight from my shoulders, cared for my family by herself many, many times and who gave me the time to write. I thank you from the bottom of my heart for motivating me, accepting the countless overhours and particularly for proof reading the entire thesis and helping me with my English. This thesis is dedicated to you. Also big thanks to my boys, Max, Arthur and Falko, for not being angry, when I was absent very often and could not spend time with you.

My special thanks go to Prof. Michael Vorländer, of course for supervising this work, but even more for giving me the opportunity to become a scientist. Since the first lecture I attended of you (Elektrotechnik für Informatiker), you became a source of motivation and inspiration for me. Thank you for your support over all these years, your friendliness, for believing in my skills and for letting me work the way, I can do it best. You made me become an acoustician. I would especially like to thank Prof. Lauri Savioja for co-supervising this thesis. The discussions with you on signal processing and numerical simulations were very motivating and encouraging for me. I highly appreciate your detailed comments and remarks, which clearly helped to improve this work. I would like to thank Prof. Kay Hameyer and Prof. Joachim Knoch for agreeing to be co-examiners in my doctoral exam.

A very big thank you goes to my friends Ramona Bomhardt and Jonas Stienen, who proof read the entire thesis and gave me their view on the document. Thank you for your time, your detailed remarks and very valuable feedback. Without you, the thesis would not have become what it is. And I promise, I will not bother you with fast convolution again.

The period of being a research assistant at the Institute of Technical Acoustics (ITA) was clearly one of the best times in my life so far. It is hard to believe, that by the time of handing in the thesis, it has been already eleven years since I started at ITA as a student worker. During this time, ITA became a second family for me and second home for me. I would like to thank my colleagues and co-workers for their friendship and support. Particularly,

Dr. Gottfried Behler for countless talks, giving me advice on every possible problem, for showing me the right direction and helping me understanding acoustical problems. I owe you a lot. I thank the past and present members of the virtual acoustics group for their teamwork, namely Dr. Dirk Schröder, Sönke Pelzer, Jonas Stienen and Lukas Aspoeck. Sincere thanks to Dr. Tobias Lentz for a great period of time as student worker for him. Without knowing how you did that, you definitely put the convolution virus in me. Thanks go to Dr. Pascal Dietrich for very enjoyable bilateral discussions on hardcore signal processing in the ITA coffee corner. We should have had more of these. Special thanks go to Rolf Kaldenbach, head of the electrical workshop, for a very relaxed and friendly collaboration over all the years. It was a joy to work together with you. Thanks also go to Uwe Schlömer and his crew of the mechanical workshop, for letting my dreams and wishes become condensed matter. Thanks go out the secretaries Ulrike Goergens and Karin Charlier for helping with countless organizational tasks. Finally, I would like to thank my students, Christoph Röttgen, Nikolas Borrel-Jensen, Jan Richter, Penelopi Ghika, Christian Sander, Jörg Seidler and Tobias Schöls for their great contributions. Particularly, I thank my student Philipp Schäfer for inspiring discussing on fast convolution and for his valuable remarks on parts of the thesis.

If ITA is my second home, then I have also a third one: The virtual reality group (VRG) of Prof. Torsten Kuhlen, running the CAVE environments at RWTH Aachen. It was an absolute privilege to work with you over all these years. Thanks to Prof. Torsten Kuhlen, Dr. Bernd Hentschel, Dominik Rausch, Sebastian Pick, Andrea Bönsch and Andreas Hamacher for the truely great cooperation and for the opportunity to work on the fantastic VR infrastructure. Building the aixCAVE acoustic system was a unforgettable experience. I thank Dr. Ingo Assenmacher for his great commitment in the long-term cooperation between VRG and ITA and for many inspiring discussions on software engineering. I learned many things from you. A person who accompanied me over many years is Dominik Rausch in the VR group. The joint achievements of VRG/ITA would not have been possible without your support, commitment and unlimited help in case of short-term problems. I award you the trophy of the most helpful person on the planet.

Three people helped me with the preparation of my oral exam. My sincere thanks go to Sergej Fischer for his efforts explaining quantum physics and material science to me, to Thorben Grosse for providing me material on renewable energy and Sandra Wienke for updating me on the cutting edge in graphic processors.

Finally, I would like to thank my friend Philipp Marla (formerly Heck) for getting me in touch with the Institute of Technical Acoustics in the first place. Without you, I would never have taken this road...

Bisher erschienene Bände der Reihe

Aachener Beiträge zur Technischen Akustik

ISSN 1866-3052

10	Sebastian Fingerhuth	Tonalness and consonance of technical sounds
		ISBN 978-3-8325-2536-1 42.00 EUR
11	Dirk Schröder	Physically Based Real-Time Auralization of Interactive Virtual Environments
		ISBN 978-3-8325-2458-6 35.00 EUR
12	Marc Aretz	Combined Wave And Ray Based Room Acoustic Simulations Of Small Rooms
		ISBN 978-3-8325-3242-0 37.00 EUR
13	Bruno Sanches Masiero	Individualized Binaural Technology. Measurement, Equalization and Subjective Evaluation
		ISBN 978-3-8325-3274-1 36.50 EUR
14	Roman Scharrer	Acoustic Field Analysis in Small Microphone Arrays
		ISBN 978-3-8325-3453-0 35.00 EUR
15	Matthias Lievens	Structure-borne Sound Sources in Buildings
		ISBN 978-3-8325-3464-6 33.00 EUR
16	Pascal Dietrich	Uncertainties in Acoustical Transfer Functions. Modeling, Measurement and Derivation of Parameters for Airborne and Structure-borne Sound
		ISBN 978-3-8325-3551-3 37.00 EUR
17	Elena Shabalina	The Propagation of Low Frequency Sound through an Audience
		ISBN 978-3-8325-3608-4 37.50 EUR
18	Xun Wang	Model Based Signal Enhancement for Impulse Response Measurement
		ISBN 978-3-8325-3630-5 34.50 EUR
19	Stefan Feistel	Modeling the Radiation of Modern Sound Reinforcement Systems in High Resolution
		ISBN 978-3-8325-3710-4 37.00 EUR
20	Frank Wefers	Partitioned convolution algorithms for real-time auralization
		ISBN 978-3-8325-3943-6 44.50 EUR

Alle erschienenen Bücher können unter der angegebenen ISBN-Nummer direkt online (http://www.logos-verlag.de) oder per Fax (030 - 42 85 10 92) beim Logos Verlag Berlin bestellt werden.